The Bay Audio Guide to

Companion Selling®

By Ira Friedman
President, Bay Audio

BayAudio
Published by Bay Audio, LLC
2007

Published by Bay Audio, LLC
14290 Danielson Street, Poway, California 92064

ISBN: 978-0-6151-5899-0

Printed in the United States of America

Contents

Companion Selling.

Why do interior designers seem to get whatever they want? Why do your clients double-check your proposals for "sanity?"

It's not their fault. It's yours.

Because you're thinking about products and services like a retailer and not like a consultant. It's because your business is geared to making a sale, and not developing an ongoing relationship, and finally, because you aren't attacking your clients with the right level of exuberance, enthusiasm, and awe.

This book lays out a new approach to running your business. It's an approach I call *Companion Selling*. Complete in scope, *Companion Selling* includes recommendations for acquiring clients, making a sale, and maintaining long-term relationships, as well as guidelines for project design, invoicing, and strategic financial considerations. *Companion Selling* is a complete way of doing business. It's not for the meek or inquisitive, but structured for managers who want better control of their clients, their expenses, and ultimately, their profits.

There are aspects of *Companion Selling* you are currently employing in your business. You'll say, "I do that," as you read through the chapters. But doing some, or even most, of the *Companion Selling* recommendations is not the same as operating like a *Companion Selling* company. You share 99% of your DNA with a chimp, yet I'm sure you are 100% different from Bonzo. And so, *Companion Selling* works best when all the strategic issues are fully incorporated in your business.

A Warning Before You Delve Further.

If any of the strategic issues laid out in *Companion Selling* don't ring true, or if any of the recommendations seem preposterous, simply put the book down and continue on your way. *Companion Selling* is a strategic plan, pure and simple. It is not the only strategic plan you can use, and it may not even be the most lucrative strategic plan for your business, in your market, with your personnel.

It is a comprehensive approach to our business as I see it today – in 2007. And based on a handful of Bay Audio dealers who learned and practice the *Companion Selling* approach, it can be a successful strategy that doubles your profit as soon as you begin implementation.

CHAPTER 1

Chantel Gets Her Way.

Chantel is an ASID card-carrying interior designer. She knows flocking when she sees it. She can tell a Louis XVI from a Chippendale. She's worked with your client on previous projects, in previous homes, and she's running the show.

So when Chantel says the plasma is going above the fireplace (which, as you know, is about three feet too high), and the speakers are going in the ceiling (which won't give your client the "movie theater" experience they asked for), you are unable to resist.

Indeed, you've been relegated to the heap of subcontractors working in the home – the drywaller, the electrician, the painter. None of these trades can get Chantel to consider their input. Chantel just doesn't care.

And since the clients trust Chantel implicitly, you're playing with fire if you think you can have a conversation with Mr. and Mrs. Homeowner counter to Chantel's recommendations.

So you relent and install a system that indeed looks great – Chantel is talented, you tell yourself – but has compromised performance. Your referral possibilities have been crushed as you imagine your client's first dinner party with AV inquisitive friends. "Boy, that doesn't sound so hot, and the plasma is too high to watch comfortably," they muse while sipping Pinot, "I wonder what their AV guy was thinking . . ."

Chantel's Secret Weapon.

Chantel has something you don't: a great relationship with the client. Chantel has proved herself in the past – she is a good designer, and she gets the job done. You, on the other hand, are a hired gun.

Not to denigrate your skills; surely you are a proficient hired gun. You may even believe your relationship with the client is "rock solid." But Chantel's is rock-solider.

What did she do to earn this level of trust?

She built a working relationship before she started the project.

This is Step 1 in *Companion Selling*.

You *must* have an honest relationship with your clients before you begin work.

Without it, you are a fantastic subcontractor. With it, you are a trusted resource.

Four Important Parts and One Clincher to Building the Perfect Working Relationship with a Prospect.

Chantel earned her place in the client's heart through a combination of factors, each of which you also must achieve.

First, the client knows Chantel earns her money by consulting, and not by selling products from inventory.

Chantel has a portfolio of work she can refer to ("You know I designed Puff Daddy's cottage") or show ("And here's a picture of P lounging in the living room. Boy did he like fur").

Chantel has a little black book filled with money-saving tips and tricks, plus a list of trusted subcontractors willing to "discount" their services for her. She shares her tips and contacts with her client.

Next, Chantel is fun and creative.

And finally, Chantel is passionately, enthusiastically, overwhelmingly in love with every project she tackles. So much in love, her enthusiasm bubbles over and infects her clients whenever she speaks.

Chantel has developed a relationship built on honesty. She shares information freely, and she's eternally enthusiastic. This is a potent combination.

The Ethan Allen-ization of Consulting.

There are ASID-certified designers working in your local Ethan Allen store. Many are quite accomplished.

And yet, when you walk into an Ethan Allen store, you do not expect to get an impartial design consultation. Because these designers are directly tied to Ethan Allen, their impartiality is gone. They may sell you a beautiful living room – but rest assured – it will be stocked with Ethan Allen products.

So when you walk into an Ethan Allen store, you assume you'll walk out with Ethan Allen furniture and you hope the designer on staff will be good enough to help you match a few fabric samples and paint colors.

Working in the store is a great gig for designers who are not interested in running their own business. With nice commissions, and built-in clientele, they earn a reasonable living. But their association with Ethan Allen has all but neutered their ability to design freely as Chantel. They've become less "creative."

On the other hand, Chantel isn't shackled to Ethan Allen. She can specify any furniture she wants. But your prospect senses that you are shackled to your brands. When you make contact with a prospective client, they believe – rightly or wrongly – you'll be selling them the couch out of inventory, regardless of their desires.

Companion Selling recognizes consultants are free to make recommendations 100% appropriate to the client. And clients put a higher value on a consultant than they do a on salesperson.

Clients like consultants.
They don't like salespeople.

Now here's the irony: Chantel limits her choices in order to maintain her sanity. She has a trusted list of suppliers, just like you do. Time after time she finds herself using the same cabinet shop, the same painter, the same drywall company. She likes buying antiques from one trusted dealer. She continually steers her clients to certain furniture brands she likes. She prefers certain flooring brands, tile companies, and faucet manufacturers. Chantel does not start with a clean slate of suppliers. Like you, she has her go-to brands.

However, the client doesn't find this a negative. The client knows Chantel starts with an open mind when it comes to arranging these elements. In fact, Chantel could steer each and every one of her clients to Ethan Allen furniture, and this wouldn't reduce her value – or make her any less of a consultant.

Said another way, Chantel, operating as a consultant, is free to recommend any solution to her clients. That's powerful. The fact that she might happen to recommend a similar solution to all her clients is irrelevant.

You, like the Ethan Allen designer, have inextricably tied yourself to product. Don't believe me? When you're on the CEDIA shuttle heading back to the airport hotel and the guy next to you says, "What brands do you carry?" you rattle off a list in seconds flat.

When Chantel's on the shuttle, returning from an ASID conference, her seat mate wouldn't think of asking what brand of couch she supports.

Your prospects sense this. They believe Chantel is impartial, and that you have an agenda. Consultants have agendas, of course, but not based on product. Chantel's agenda is to design a beautiful room for her client, and her client knows this. They aren't so sure of your agenda.

The Trading Line.

Draw an imaginary line in the sand. Let's call this the Trading Line. The Trading Line represents a transactional event. You're on one side of the line. You are the prospective client. On the other side are the people from whom you are buying.

Where would you place the designer from the Ethan Allen store? On the other side of the line, of course, because this designer is about to sell you something from inventory when you walk into the store. And even though you're interested in an Ethan Allen solution, this is a transactional relationship.

You'd place a car salesperson next to the Ethan Allen designer. And the sales kid from Circuit City's camera department.

Now imagine you're about to climb to Everest Base Camp in the Himalayas. Tenzing Norgay, your trusted sherpa will lead the way. On which side of the line would you place Tenzing? Hopefully not with the car dealer. That could be dangerous.

Where would you place your family doctor – the one who revels in hammering away at your kneecap with a rubber mallet? How about your CPA? Your personal trainer? Your baby sitter?

These trusted individuals are on your side of the line, even though you have transactional relationships with each one (you do pay them, of course). If you are not able to develop a level of trust with these people – the ones responsible for your heath, your finances, your kids – then who would you trust?

Now place Chantel on the grid. She's right there next to Dr. Hammer. Another trusted member of your personal posse. Because Chantel wouldn't – nay, couldn't – design your living room without you loving it. She works for you.

Now place yourself back in your own AV persona. On which side of the line do you think you should be placed? Of course, next to Chantel. But where does the prospective client actually place you? Oftentimes next to the car dealer. Sorry. You don't get to jump the line until you develop a meaningful relationship. Until you stop selling and start caring.

Gaining Trust.

Doctors gain immediate trust because they've earned it through schooling. Their degree confers trust. You don't have to spend an inordinate amount of time trusting your doctor will make decisions good for you.

AV salespeople don't get a degree in AV-ology, so it's up to you to build your credentials quickly with your prospect. You've got to prove you're no AV quack. To prove your value, start by telling your client about your storied past accomplishments. ("I designed Puff Daddy's house in Boca.") And then show your prospect the quality of your work, ("Boy, did he like sub-woofers.")

Showing your abilities is vital in gaining your client's trust. The more impressive your portfolio (and client list), the better impression you'll make.

Notice I didn't say that you have to play a demo. You don't have to pass your client a touch panel. Once you do, you've become the Ethan Allen designer – a specifier. There's plenty of time to talk about product later.

ACTION ITEM

Establish your qualifications.
Create a portfolio of work. Print a referral list.

Chantel's Little Black Book.

Talk at length with any good designer, and you'll find she has a collection of "insider" tips and tricks, like the stucco guy from Sicily, or the drywaller who does touch-up work at the Guggenheim. Designers, because they are consultants and not tradesmen, must assemble a cadre of great resources in order to create their vision. What is a great idea if it can't be executed properly?

So when Chantel meets a prospective client, she quickly brings up her secret list of tradesmen.

> *"I know you want to use stucco in this room, and I know this great Italian stucco artist – he's to die for – straight off the island of Sicily. He only does work for the best clients, and I use him for my special projects. He always gives me preferential treatment, and discounted pricing. We'll absolutely use him on this project!"*

By talking up the stucco artist, Chantel proves she's an expert, and since she "owns" this resource, the client understands she only has access to the stucco guy if she uses Chantel.

You need a Secrets List, too. One that includes the same tradespeople, including drywallers, stucco guys, cabinetmakers – even specialists, like faux painters and lighting designers. Develop and hone this list. Tell these tradespeople you'll be referring them. Make sure they understand the AV business and its peculiarities.

Chantel also has a list of secret money-saving tips, like how to get European fabrics for half price and have them sewn into curtains by a local seamstress. And how to use inexpensive industrial materials – like cork sheeting, aluminum, chain link – in residential applications. As Chantel talks with her prospective client, she pulls out one or two of these money-saving secrets and offers them up for consideration.

What is Chantel doing? First, she's proving her resourcefulness. The client wouldn't know to use industrial cork sheeting as flooring, so Chantel must be smart. Second, she's spending the client's money as if it were her own.

> *"I like to take entrance door designs and send them to a great little woodshop in Mexico. They carve these beautiful doors and ship them to the States at a fraction of the cost. We can save a ton of money this way."*

Third, she's showing her creativity. By thinking out of the

box, and looking for creative ways to appoint the home, Chantel is gaining her client's trust.

What are your secrets? Write them down, hone them, make them a part of your company's presentation. Here's an example:

> *"I know we should be isolating the sound from this surround system because the baby's room is so close. I know a way we can get industrial sound dampening sheets and place them behind the drywall. It's a fraction of the cost of the fancy consumer product, and works even better."*

Parity Pricing is a "Secret."

Add examples of parity pricing into your Secrets List. For example, prospects have an inkling as to how much a 50-inch plasma costs. So let them know you'll be supplying plasmas at parity pricing with the market – but remind them that your sets are significantly better. For example, you can offer a commercial version of a residential set:

> *"We like to use the commercial version of this plasma. It's priced about 10% below the residential version, comes with a longer warranty, and performs exactly the same. It's a great way to save a few dollars."*

Don't be afraid to acknowledge pricing on competitive items like video. However, make sure you have a clever way to provide greater value through your Secrets List. We'll discuss the details of pricing in ***Chapter 5***.

ACTION ITEM

Write down your Secrets List. Include:

1. Tradespeople names and phone numbers.

2. Unique money-saving tips.

Using "What If" Questions.

When Chantel meets her prospective client, she asks a lot of "What If" questions.

> *"I know you want this room to be brighter and sunnier than it is now.* ***What if*** *we knocked through this wall and added a window right here? That would look great!"*

Or,

> *"****What if*** *we added extra lighting fixtures to bathe the room in light. I know a great lighting designer who can make this room absolutely shine."*

"What If" questions get the prospect involved with the design process. They elicit a response – and begin to focus the client on a narrower set of options. "What If" questions, far from being weak, strengthen the bond between Chantel and her client. She has to ask creative, fun, intelligent "What If" questions, of course, but forcing the client to think beyond their limited view opens the design up to wonderful possibilities.

What are your "What If" questions when you meet a prospect? Do you say:

> *"I know you want to watch TV without a glare on the set during the day. What if we had shade control in this room, so when the TV is on, the shades partially close. There'll be plenty of light, but we can remove the glare."*

"What If" questions are thought provoking. They focus your prospect's attention on possibilities, and away from budget. And they can be used at any time during the course of the project – even as you are coming to completion.

> *"Now that we've mounted the plasma,* ***what if*** *we changed the bezel to match the lighting fixtures? That could look really good."*

"What If" questions establish a strong relationship with your prospect. Use them generously.

Now Here's the Kicker.

Chantel is the most enthusiastic subcontractor working with the client. The architect isn't. The builder certainly isn't. And the plumber, electrician, and flooring guys? They're just doing their job.

But the designer is all bubbles and fun and energy. The designer trades in "what's possible," whereas everyone else is selling "what's practical."

There are no rules in the design world. There is no "not-to-exceed" budget. The scope of the design is only limited by imagination and cash flow. Only the client knows how much cash flow there is. The designer is blind to that reality. She just wants to create exquisite art with the client.

Everything is exciting in the world of design. Small projects with tiny budgets are exciting because they present an artistic challenge. Large projects are exciting because they're comprehensive. Even the most distinguished and internationally recognized designers will accept smallish projects, because they make money selling their time, not the couch.

AV Dealers, on the other hand, get really excited about large projects because they see the dollar signs associated with big-ticket parts. Fair enough. But the prospective client senses that – and is completely turned off by the ho-hum attitude their AV sales guys possesses when faced with a smallish budget.

Your job is to be enthusiastic with your prospect regardless of the budget they initially offer. You should be enthusiastic about theaters, automation systems, lighting control, and outdoor music systems. The enthusiasm is contagious. You can deal with budget constraints and issues later, as we'll see in ***Chapter 7:*** *Honoring the Budget*.

CHAPTER 2

Using Inclusive Language.

Listen to an interior designer speak with a prospective client and you'll hear different vocabulary than you use. Designers use inclusive language, like "us" and "we" when talking about a project.

They'll say, "This is going to be a great living room. We're going to have such fun designing it."

Or you'll hear them say, "What kind of budget are we working with?"

Inclusive language is a subtle way of telling the prospect you're on their side of the line. I don't believe designers use inclusive language as a manipulative trick. Instead, they use language that reflects their position with the client. As a partner.

You should begin to use inclusive language with your prospective clients. When you make a conscious effort to do so, you'll find yourself moving closer to your client, and inevitably you'll begin treating them like a team member, not like a customer.

Words to Avoid.

Try not to say:

You
My company

Instead, say:

We
Us

When talking about the prospective client's project:

"Tell me more about the distributed audio system

we'll be designing together."

"Putting a completely hidden surround system in your bedroom will be fun. Where should we put the furniture that houses the plasma?"

Now, it's impractical to never use the word "you," but switching to inclusive language, as shown above, brings you and your prospect closer.

When talking about your own company:

"AV Solutions has perhaps the best talent in the area. The company is known for its engineering and design skills. We'll take advantage of the company's talents with this project – I think we'll be extremely pleased with the outcome."

Notice you begin to present your company at arm's length. You do this to reinforce your standing as a consultant and ombudsman for your prospect. By slightly distancing yourself from your company, you are better able to speak of its strengths (and perhaps weaknesses) while strengthening your ties with the prospect. You're selling yourself during these initial stages. And unlike the designer working at Ethan Allen, you have to show you are somewhat independent – even if it seems like artifice – in order to establish your value as a consultant.

When talking about other subcontractors:

"I know this great painter who can paint a faux finish on these in-wall speakers. We'll have a hard time seeing where the speakers are once he's done."

"We should think about having a custom cabinet built for this room to hide the plasma and the lift. I've worked with this wonderful cabinetmaker for

several years, and he does great work. We'll be happy with his turnaround time and discounted pricing."

Make Inclusivity Your Goal.

Try to insert yourself whenever possible into the prospective client's decision-making process. "What if we had a custom cabinet built? That would solve a lot of the aesthetic issues we're facing."

Be willing to face dejection from being inclusive. Prospects may answer, "No, I don't like that idea." They're feeling no reason to be inclusive with you. However, that shouldn't change your approach. At some point, you'll hear your prospect start to use inclusive language as well. That's when you know you're working together as a team.

CHAPTER 3

Using Your Prospect's Input.

Like the trusted trail guide, you lead through the forest. But you have to answer questions and be open to suggestions.

There are certain aspects of the project you decide:

1. Products
2. Process

There are aspects of the project you may allow your client to help design:

1. Aesthetic issues
2. Interface

Avoid the Puma.

You choose the products. You choose how they're installed. That's not up for discussion. If the prospect has product suggestions, steer them to your choices. Why is this important? Because you cannot control the project unless you control the products. It's that simple.

Like the trail guide, you choose the path. If a hiker asks to go up the north side of the hill, and you know it's the home of a man-eating puma, you say, "No, that's the home of a man-eating puma. We'll take the south side." The same is true with your AV design. When a prospect asks if he can have a system design you don't favor – or products you won't carry, tell him "no," explain why you've made the choices you have, and move on. There's no reason to dwell on the danger the puma poses. It's a puma.

Letting a prospective client choose products makes you no more than a glorified valet. You lose control of performance, of interoperability, and of success. Letting the client dictate even one product you don't carry is a mistake. Avoid the puma.

Most clients don't want to choose the products. That's what they hired you for. If they enjoy shopping for furniture, if they keep swatch books, if they tour model homes on weekends, then they don't need Chantel. If your clients spend time browsing stereo magazines, if they build ham radios for fun, if they spend weekends in Circuit City, they aren't your clients. They know where the pumas are, and don't need your guidance.

But your clients, the clients who need guidance, want to make two choices: the first is with whom to work. The second is aesthetic. Clients want to choose their system interface – and determine how it looks in the home.

No negotiating allowed:
You choose the product and process.

Let the Client Make Aesthetic Decisions.

Every aesthetic decision should be one the prospective client has an opportunity to make. Since you're operating like a designer, aesthetic decisions are numerous and important – and become the focus of the project.

The more energy you put into the aesthetic decision-making process, the easier it will be to get the prospect enthused. They don't know HDMI from RG40, and they don't want to know. But they do know if they want the big plasma TV over the fireplace or tucked into custom cabinetry. They know what color faceplates they should have and where the keypads should be located.

Even when aesthetic decisions compromise performance, you should let the prospect know the trade-offs, but support them.

Focusing on aesthetics allows the prospect to avoid these embarrassing conversations:

"I don't really watch that much TV, so an expensive set isn't important to me."

Many clients don't believe they watch as much TV as they do. And they certainly won't tell you about it. So trying to sell a high-end TV on performance may backfire.

Talk about the design of the set. Talk about fit and finish, and how the TV becomes a part of the room. Tell your prospect how good it will look on the wall. And of course, mention the performance of your higher-end offerings, so when they do watch TV, it'll be crystal clear. People are confident they can make a visual decision, and they understand that a clearer picture is a superior one. So sell the aesthetic benefit of the picture.

Clients say: "I can't hear the difference."

Of course they can hear the difference, but it will take too long for you to prove the point. Instead, tell them they will hear the difference, and they'll appreciate the sonic quality of the system – after it's installed. Prospects say they can't hear the difference. They'll tell you they aren't audiophiles. They do this to protect themselves from buying expensive speakers.

Remember the Trading Line from ***Chapter 1***. Prospects put you on the other side of the line until you prove your value. Disavowing a preference for sonic quality is a defensive maneuver. Unlike visual acuity, most people do not believe they can hear a difference.

But you know better. Hearing, like all senses, can be fine-tuned with practice. Your client knows a good cup of coffee. But would they have known a good cup of coffee 30 years ago? Had you tried to sell "better tasting, more expensive coffee" in 1970, you would have struggled to make a sale. Starbucks didn't just introduce better tasting coffee, they had to re-orient the customer's entire coffee drinking expectations. You don't have the time, nor the budget to fight that battle.

Instead, focus on the aesthetics of speaker design, size, and placement. You can talk about hiding speakers in furniture or in the walls. You can show how speakers can be placed outside so they can't be seen. Selling the aesthetic decision will lead your prospects to buy better-quality speakers. Once installed, they'll learn to appreciate better sound.

These are aesthetic discussions that lead you to sell higher quality, better-looking speakers. Because even if they can't hear the difference, they can see it.

Let your prospect choose the look, the colors, and the placement of the products you've chosen for them.

Let Your Client Dictate the Functionality of the System.

Functionality refers to the user interface – how your client controls the system, where the controls reside, and how easy it is to navigate the controls.

Involve your prospects in every detail related to functionality. Show them how you design systems by presenting a live touch panel or remote. Show them a portfolio of your work with rack-mounted electronics in out-of-the-way places.

Explain the options your prospect has in choosing what they can control, and to what degree. Each presented option should be asked as a question, such as:

> *"We can simplify this system so anyone in your family – even guests – can operate it intuitively. There are plenty of options we can consider. What would you like to achieve?"*

Remember the "lesson of the iPod": it is a successful product because it's intuitive and beautiful. You can propose intuitive and beautiful systems as well. Let your prospects determine the level of elegance and functionality they want in their systems. You'll find that if they drive these decisions your projects will grow exponentially in scope.

You may encounter objections like these:

"I hardly listen to music, so I don't need an audiophile system."

Here's an opportunity to talk about functionality. Explain how easy it is to use an automated distributed audio system – so easy, in fact, they'll find themselves listening to more music throughout the day. Tell them how the iPod, with its ease of use, rekindled the music industry. People who had never considered surrounding themselves with music now rely on their iPod on a daily basis.

We all know music in the house makes the home feel better. This is how you turn your presentation to functionality and away from "audiophile."

"It's not hard to hit a light switch. Why do I need a lighting control system?"

Selling functionality requires a dose of story telling. Here's a possible answer:

> *"Since your husband travels so often, perhaps we could put a lighting control system in. What if you could shut off all the lights in the house at night without walking around – right from your bedroom? That's really convenient. What if the kids could turn on a hallway lighting pathway so they can get to the bathroom at night? What if you could turn the outside lights on when you heard a car pull up the driveway? It might seem extravagant, but my clients always thank me after the system has been installed."*

Let your clients focus on system functionality and interface.

CHAPTER 4

Developing the Performance List.

A Performance List contains all the prospect's wants and needs. Your prospect says they want a surround system in the bedroom hidden from view when not in use, and you describe the perfect solution.

Key to your description is the knowledge that your prospects have a limited understanding of the aesthetic, control, and performance possibilities available to them. They might be thinking the plasma hides in a piece of furniture, but you describe how the control system activates a lift, dims the lights, closes the shades, turns on the audio system, and lets them choose their source.

Paint the picture using the same vocabulary Chantel uses. Talk about how beautiful the furniture will look in the bedroom. How the screen rises silently and swivels into position. Tell them the speakers are mounted into the ceiling – and painted to match. How the electronics are discretely rack-mounted in the closet, and how the lights can be programmed to different settings.

While setting the scene, use those handy "What If" questions:

> "***What if*** *the shades automatically closed part-way when you turned the system on? Wouldn't that be something?"*

Use Your Secrets List.

> *"I'll call my friend Bill the cabinetmaker. I've worked with him for years – just recently on the house that made the cover of* **San Diego Home and Garden** *– we should have him come down here and look at*

> *the bedroom. He has great artistic flair – and can build the perfect piece to house the plasma TV. Plus, he'll give us 'to-the-trade' pricing."*

Use inclusive language, and keep it enthusiastic:

> *"This could be a really nice retreat for you and your husband. We need to think about making it beautiful, since we don't want to see the gear overpower the room. And it's got to be intuitive, so you can easily switch from CNN in the morning to a DVD in the evening."*

Expose Your Prospect to Our Industry.

Clients don't know much about the AV world. It's your job to educate them. The best way to educate is to ask those "What If" questions. "What If" questions pose a condition – and supply a solution; that's what makes them powerful.

You could educate your prospect by laying out options. And that would be wrong. This is not an exciting exchange:

> *"We could do lighting control. Is that something you're interested in?"*

Don't laugh. Listen carefully to most sales presentations, and that's how options are discussed. Like this: "We got bananas. You want bananas?"

A "What If" question creates a scene.

> *"**What if** we had a lighting control system, so when you turn on the TV, it automatically adjusts the lights in the room on a dimmer? We could have the front half of the bedroom get darker, to make the picture stand out, but keep the lights over the bed a bit brighter."*

Clients can agree, disagree, or offer up another option. In any event, you get them thinking about lighting control and its possibilities. They might even ask, "How much is that going to add?" to which you offer your best Ballpark answer. We'll review the Ballparking process in depth in ***Chapter 5***.

Reviewing the Performance List.

The Performance List is a story. It is not a line-item listing of parts, pieces, and prices. It's the first step to developing shared language both you and the client can understand.

Each significant constituent of the system gets its own Performance List. There should be one for each discipline – automation, lighting, audio/video distribution, theater. You might find it helpful to have Performance Lists for each master zone – for example, upstairs, downstairs, and the backyard.

Performance Lists are rough ideas that describe the functionality, aesthetics, and performance levels of a system. In some cases, they will overlap – the outdoor system has a Performance List which describes the audio and video experience on the patio and at the pool. The whole-house automation system has its own Performance List, which includes control of the patio and pool areas.

No reason to get too hung up on Performance List details. You're simply writing a story with prospective clients. A story you and they can repeat with some level of clarity.

Here's a Performance List for our bedroom surround system:

> *"You'll be able to wake up in the morning, press a button on a hand-held remote control, and have CNN appear on the plasma screen that automatically lifts out of a piece of custom furniture. Speakers in the ceiling will fill the room. Turning the TV off also returns the plasma into the cabinetry and shuts off*

> *the electronics hidden in a rack in your closet. When you want to watch a movie, just press the MOVIE button on the remote. The shades will close, the plasma will lift from the cabinet, the lights will dim, and a screen will appear on the remote, giving you access to a media server hidden in your closet. We'll pre-load the server with 100 movies. One button turns the system off."*

That's a Performance List. It contains the important aspects of the system. Of course, clients have their own special details they'll add to the List – like the client who insists on having wireless headphones for her spouse. Or a video distribution system. You'll uncover these details by asking the "What If" questions.

You'll know when your Performance List is complete when your client can confidently repeat the List back to you. And here's the best part of Performance Lists: you've given your client the tools she needs to describe her system to others. As she retells the Performance List story to family and friends, she'll become more confident, and perhaps adventurous, with the system. She'll become a proponent.

Build a Performance List carefully.
It's your road map.

CHAPTER 5

Ballparking.

Once you've developed a Performance List and you're able to recite it back to your prospect, it's time to start Ballparking. Ballparking is your way of immediately sharing pricing information. It's your rough estimate for each constituent and discipline of the project.

Ballparking is the not the same as estimating. It is a way for you to test your prospect's understanding early on in the process. *Companion Selling* treats estimating and proposal writing differently from traditional selling approaches. The *Companion Selling* method uses a three-step process to determine the selling price for a project.

Step 1: Ballparking

Step 2: Honoring the Budget, discussed in ***Chapter 6***

Step 3: Engineering and documentation, discussed in ***Chapter 7***

There are several approaches to Ballparking. Pick the one that suits your personality.

1. *Compare this Performance List to a previous project.*
   "We worked on a bedroom surround system quite like this a few months ago. All said and done, it was about $100k and took two months to complete. Is that what you were expecting?"
2. *Be blunt.*
   "We're looking at about $100k in engineering and product, and about two months for completion."

3. *Soft sell.*
   "The way we've described this bedroom system, with all of the automation and aesthetic considerations, we're looking at $100k or so by the time it's completed. I figure it'll take about two months with the custom cabinetry and shade control pacing the project."

Stay On Your Prospect's Side of the Line.

To remain a trusted advisor, it's imperative you share information with your prospective client as it occurs to you. You are not permitted to withhold information your client needs in order to make a good decision.

Share information immediately if it has a material effect on the project.

If you fall into the trap of holding back information until you are absolutely comfortable in sharing, it's probably too late – the prospect knows you're holding back and the frustration level rises.

How often have you been in a selling situation where the prospective client asked you, "So how much is this going to cost?" and you answered, "I don't know. I need to go back to the office and put together an estimate."

Remember how frustrated the prospect was? He probably begged you for a Ballpark estimate. But you countered by saying something condescending, like:

> *"I'd rather not give you the wrong price now, because if it's too low, I'll just have to revise the pricing later, and if it's too high, I'd be taking advantage of you. Let me go back to the office and work up an estimate."*

Boy, is that frustrating. Don't you hate it when you ask a salesperson for an estimate, and he pushes you off? You're thinking one of two things:

1. The salesperson isn't smart enough to know what he sells, or;

2. The salesperson is arrogant.

In either case, you're not feeling good about the sales process. You begin limiting the amount of information you share with the salesperson. You decide the conversation won't get much further until you see an estimate, and that might take several days. Your enthusiasm drops, and the process becomes purely transactional – and no longer fun.

Chantel Loves to Ballpark.

The interior designer, on the other hand, loves to throw out numbers. She's Ballparking all the time.

> *"I'm sure we can get a great hutch for under $5,000. I know a wonderful antiques dealer who sells only to the trade. He can get us something absolutely charming."*

An interior designer never says, "I don't know how much it'll be to furnish this room. I need to go back to my office to work up an estimate." If clients heard this, they'd think the designer was up to something nefarious. What could possibly be in that office that's so magically wonderful? What information is missing? Why is this person avoiding me?

Ballparking Proves Your Abilities.

You can only Ballpark successfully if you understand your business inside and out. And the only way to understand your business is to study it.

If your mom asked you to quote a distributed audio system,

you'd be able to rattle off a set of numbers during your conversation with her. You wouldn't say, "Hey, mom, I know what you want, but I have to go back to the office to work up some numbers."

Why can you Ballpark so effortlessly with your mom – and not with your prospective client? Because you're afraid of saying the wrong thing to the prospect. And that would be bad, right? Wrong.

You can make mistakes with people who trust you – as long as you can fix those mistakes rapidly. You can't make mistakes with people who don't trust you, because they believe the mistake becomes the basis of the relationship. "I met this guy, and he estimated way too low. How can I trust him?"

Making a significant mistake during the Ballparking stage is a problem. So don't make a mistake. Learn your business. Begin to develop packages that are 90% standard, and 10% client-specific. Test your Ballparking abilities by doing rough estimates – and then double-check your math with a full-on proposal.

You aren't ready to Ballpark
until you can get the numbers right.

To Ballpark successfully, you'll need to develop a good set of estimates. We'll discuss the steps of developing these estimates in ***Chapter 6: Building the Backbone***. Then you can revisit this Ballparking chapter to tie up any loose ends.

CHAPTER 6

Building the Backbone.

Many AV dealers struggle to build "standard" systems they can present to clients. They believe these standard offerings make the selling process easier. They intuitively know standards reduce install time, programming, and follow-up. These dealers often abandon their standard systems after a short period of salesperson grousing.

What went wrong? Although a standard system is appealing to the seller, it isn't appealing to the buyer. Think back to our Trading Line. A salesperson presenting a standard system is no different than the car salesperson presenting a vehicle. "Here's our offering. It costs $50k. Would you like one?"

Clients want a more interactive selling process. How interactive? Let's say 10%. They want to decide the aesthetic and functional aspects of the system. Typically referred to as the "interface," these are the elements of a system the client sees, touches, and interacts with. The size and feel of the touch panel. The plasma on a lift. The physical placement of in-wall speakers.

You can standardize 90% of the system's functionality. Let's call this the Backbone. And you can customize 10% of the system. We'll call this the Interface. The control system is a Backbone decision. Choosing the touchpad and deciding on the programming routine is an interface decision. You control the Backbone, and the client helps define the interface.

In this manner, you achieve 90% of the benefit of standardization, and the client achieves 100% of the interface control. So what are the benefits of standardization?

1. You know your product and labor costs.
2. You can set your pricing with confidence.

3. Your labor efficiency improves.
4. You can leverage your programming across several projects.
5. Your client's system is more robust.
6. Your client's system has greater performance per dollar spent.
7. You earn more money.

Know Your Numbers.

Ballparking may be the most critical aspect of your selling process. All you know about your company, about your products, about your process, has been tested. You must do your research before Ballparking, and gather a deep understanding of your labor and product costs.

Backboning:

Create several Backbone systems. Base these on past projects. Backbone systems are fundamental in nature, with placeholders for actual products. Their functionality is 90% complete. Here's a typical Backbone system for a small surround.

Surround Backbone System: $25,000

Component/Function	Price
Source 1	$500
Source 2	$500
Source 3	$1,000
Receiver	$2,000
Display	$5,000
Speakers	$5,000
Rack, cables, etc.	$1,000
Control	$2,000
Documentation	$1,000
Install Labor	$5,000
Programming	$2,000
Total	$25,000

If you look at the history of all the bedroom surround projects you've completed, you may find that they fall into the $25k range. Sure, there are differences created by display choices, levels of automation, quality of speakers, but your analysis shows that the core of your typical system is around $25k.

That's good information. You can use that average to create a bedroom surround Backbone price. Now you know it's about $25k to assemble the core components of a bedroom surround system.

When you're creating a Performance List with your prospects, mentally refer to your Backbone system. The Backbone is the least a system will cost. As you build their Performance List and add functionality, you increase the price of the system accordingly. Like so:

Surround Backbone System + additions: $30,000

Component/Function	**Price**
Backbone System	$25,000
Better display	+$2,000
Lift	+$2,000
Coordinate with designer	+$1,000
Total	$30,000

Create Backbone systems for all disciplines, including automation, lighting control, stand-alone theaters, and outdoor systems. Developing backbones allows you to think on the fly when making presentations. Backbones give you the rough estimate needed without having to go back to the office.

Key to Backboning is capturing your anticipated labor costs. All your labor costs. So when you build your Backbone spreadsheets, add labor hours in for documentation and design. Add labor for anticipated meetings with clients, GCs, and subcontractors.

After adding in your anticipated labor costs, you may find your Backbone systems are priced higher than you are used to. Surprise. You haven't been charging enough. *Companion Selling*, when handled properly, allows you to charge appropriately for your labor.

ACTION ITEM

Create a spreadsheet of Backbone systems. Use rough numbers for products and capture all labor associated with the project.

Handling additions:

Each time you build a Performance List with a prospect, you start with your Backbone system. As you discuss functionality, you add extras to your Backbone. Which means you'll need to know the incremental cost of popular additions by memory. But as with the Backboning process, you will work with round numbers as placeholders – not actual numbers for actual products.

It's hard for many salespeople (especially those with an engineering mind-set) to think in terms of round numbers and placeholders. These salespeople believe any number they quote that deviates from the actual number is a lie.

It's not a lie to say, "Systems like this are about $25,000," during the Ballparking stage. Prospects aren't looking for a firm quote this early in the process – and become wary if one is provided. The Ethan Allen designer knows the retail price of a couch. Chantel knows a nice couch sells for "about $3,000." Both are telling the truth.

ACTION ITEM

Learn and remember the incremental costs of popular additions. You will need these at your fingertips when Ballparking.

Setting Appropriate Labor Charges.

Once you begin Backboning – and I recommend you engage the services of your CPA when developing your spreadsheet – you may find you have been undercharging for labor. (See Appendix A and B for an analysis of labor efficiency.)

When you begin using *Companion Selling* you will capture more labor hours. As you capture more hours you can reduce your labor rates. There's a full discussion of this point in ***Chapter 10: Modified T&M***. So even though you assume sufficient labor *hours* when you go through the Backbone process, you will balance this against a lower selling price per hour. See ***Chapter 10*** to determine your selling price for labor.

ACTION ITEM

Determine labor rates before setting your Backbone prices. See Chapter 10 and Appendices A & B for more information.

Ballparking is Not an Estimating Exercise; It's a Selling Exercise.

Let's review the Ballparking process. First, you Ballpark after you've developed a Performance List with your prospect. The Ballpark number is thrown out to establish a value for the project.

You're able to Ballpark quickly because you've done your homework. You have a list of Backbone systems, and a good idea of popular add-ons. You build a Ballpark price by starting with a Backbone system, and modifying it to match the Performance List.

There's a good chance your Ballpark price is higher than a quote your prospect may have received from a competitor. So what! Most likely, your competitor did not develop an exhaustive Performance List and has left out valuable parts of the system.

Your prospects know this, because the system they developed with you, with their own Performance List, is most likely functionally different than the one the competitor quoted.

Should your prospect tell you, "That sounds high. I got a quote from another company, and it was a lot lower," that's your cue to answer like this:

> *"Well, then let's review our Performance List together and see what the differences are. It's likely the other company has quoted a system different from the one we've developed together."*

That's when you begin to review the Performance List and reveal your Ballparking technique. Tell the prospect:

> *"The TV, lift, and electronics in our system run about $13k. The speakers, installation, rack, and supervision are about $13k. The control system and programming run about $4k. That's how we get $30k."*

What's happening here is addressed further in the next chapter. The prospects have a budget in mind – a budget established by a competitor's quote. They'll compare your offerings to their "budget" in order to establish value.

Companion Selling relies on an open exchange between you and the prospect, with plenty of give-and-take when setting the budget, so don't let a competitor's budget throw you. It's just a number and has nothing to do with your ability to sell a great project to this prospect.

Ballparking is not your "not-to-exceed" price. It's not your estimate. It's not a quote. Ballparking is a way to establish a rough value for the Performance List.

When Chantel is working with a client, she Ballparks, too. She says, "We're looking at $10k or so to redo the kids' bedroom." That's a Ballpark figure. Chantel isn't guaranteeing the room will be $10k. She's just preparing her prospect to spend *at least* $10k for the design. Because Chantel developed a strong working relationship with her prospect, the importance of the Ballparking process is diminished. The Ballpark figure is so insignificant, the prospect shifts her focus away from the Ballpark figure – and toward the excitement of acquiring each element of the kids' bedroom.

Ballparking is Important. The Actual Price is Not.

You have to Ballpark to make *Companion Selling* successful, but the Ballpark figure you toss out isn't critical. Don't put too much pressure on yourself to assemble the most accurate Ballpark figure. There's a wide tolerance for error – about 10% in either direction. So if you come up with a Ballpark figure of $30k, that covers the range from $24k to $36k. You have plenty of time to fix the Ballpark figure, as we'll discuss in later chapters.

CHAPTER 7

Honoring the Budget.

Companion Selling works best when you respect your client. So the first time your prospective client exposes himself and his vulnerabilities, your response is critical.

Once you've offered a Ballpark estimate for a project (say $100k for the bedroom surround system), your prospective client will respond. How you handle this response determines your selling success.

You say:

> *"We're in the $100k ballpark for a system with this degree of automation, aesthetics, and performance. What is your budget?"*

Asking your prospect what he has budgeted for a system exposes him to the greatest vulnerability in your early negotiations. Should he answer too low, he's afraid to be ridiculed. Too high, and he's afraid you see him as an easy mark.

So the average prospect throws out a number below their spending capability, but high enough to keep you interested in pursuing the conversation.

Companion Selling requires that you learn the client's budget as soon as possible and immediately respond in one of two ways.

When the Budget is Ridiculously Low.

If you Ballpark $100k and the client offers a budget of $25k, you honor the budget by saying,

> *"That sounds great. And while you (note the non-inclusive language) can build an adequate system*

> *for this price, I'm afraid the systems we design are four times that price. They're fully engineered and their operation is automated and intuitive – kind of like your iPod. However, I can refer you to an installation company. They can assemble a rudimentary system to fit your budget."*

Now the prospect will have one of three responses. If he truly can't afford your services he might ask for the referral or politely disengage from the conversation. This is your signal to leave.

However, your prospect might have offered a low budget out of ignorance, at which point you've opened the door for further discussion. "$100k? We never thought it would be so much. Why is it so expensive?" he says. Use this question to review your Performance List.

Alternatively, your prospect may have a competitive quote in his hand, and he's using that quote as his budget. That's your signal to review the Performance List, and confirm if the competitor has quoted a system with the same level of functionality. Chances are, the competitor has missed a few points – precisely because they didn't develop a Performance List. As you point out the discrepancies, you'll reinforce your role as a consultant, and your competitor's role as a specifier.

Balance the Budget Against the Performance List.

Because you established a Performance List prior to Ballparking, you have a basis for your valuation. So break down the constituent parts of the system, assigning a price to each, and re-present the budget.

> *"We're at $100k because of the various automation aspects we talked about earlier. The cabinet is about $20k, the lift and plasma another $15k. Shade and lighting control will run about $20k. We talked*

> *about hiding the speakers in the walls, and possibly putting a subwoofer in the closet. That runs in the $10k region. Add the electronics in the closet, the programming, the installation labor, drywall repair and final painting, and you've got another $30k to finish it off. Altogether we're in the $90 to $100k range."*

Now honor their budget by showing what can be removed from the system, while still retaining some integrity.

> *"So what part of the system should we remove first – perhaps the shade and lighting control? That'll bring us down to $80k. How do you feel about that budget?"*

And so on.

You may continue to reduce the functionality of the system by removing building blocks, but you cannot reduce the Ballpark pricing by swapping out less expensive components. For example, if your Ballpark assumes a $5,000 plasma, don't say, "We can take a few thousand out of the system by using a cheaper plasma TV."

Addressing individual components is called pricing. What you're doing is called Ballparking, and the only items under discussion at this point in the relationship are the fundamentals: lighting, shade control, automation, audio and video.

At this point in the relationship, you are still establishing trust. You're still finding out if the client is right for you. It's premature to begin pricing discussions.

As you balance the budget against the Performance List, you'll find that either the prospective client wants you to add back in the items you've removed, or she can live without them. Either way, you're still fine-tuning the Performance List and matching it to your client's budget.

Why the Client's Budget is Low.

Should you walk into a BMW dealership and ask to drive a 750Li Sedan, you'd expect the car to cost about $75k. So you're not surprised when the dealer tells you what the monthly payments are. You wouldn't go into the dealership and say, "I'd like to buy the 750Li, and my budget, kind sir, is $50k."

The same is not true in the AV world. Your prospective client has limited knowledge of electronic and labor prices. There is no price sheet he can consult. When he establishes a budget, he's doing so without reference.

By developing a Performance List, and breaking it down to fundamental building blocks, you educate your client. Sure, he can double-check your pricing after you leave, but if you've done a credible job, if you've relied on your Secrets List, if you've proposed a high-value system, then your prospect will find your Ballparking to be fair.

Don't be surprised when clients say, "I never knew lighting control was so expensive." Of course they didn't know. They had no way of knowing. But that doesn't mean they don't want, and can't afford lighting control. It means they're looking to you to establish a value benchmark for lighting, and explain how the pricing is derived.

In our example, the prospect has decided to forego the lighting and shade control, and stick with the rest of the components. Together, you agree to an $80k budget.

Budget is Not Ability.

A prospect's budget has little relationship to his ability to buy. It's just a number concocted to represent the value he's assigned to your services. Should you increase the value of your services through education, his budget may very well go up. But if you add no value, the budget stays fixed in stone.

Companion Selling requires you to honor the prospective client's budget not because it's the final price for the system, but because it represents the client's ability – and not yours – to assign a value to a system. Accepting the budget is your way of being respectful. Using the Performance List is your tool to either reduce the functionality of the system to meet the budget, or to increase the budget as they decide to add more value to their system.

In both situations, you've removed the antagonistic "us vs. them" relationship and maintained your status as a consultant, and not an old-timey salesguy.

There is no fixed "Budget."
There's only "Value For Performance."

CHAPTER 8

Creating Desire by Being Creative.

At this point, you've discussed the project with your prospect. You've shared some of your Secrets List. You've developed a Performance List, you've Ballparked, and you've established a budget the client is comfortable with. Now it's time to build desire.

Companion Selling is a shared process. You are the trail leader walking through the forest, and your prospect is following behind. You're both on the trail –you're both sensing the beauty of nature – but *you* are clearly the leader.

The best trail guides walk the path as if it were their first time – with wonderment and awe for the beauty of nature. This enthusiasm is contagious, and they want their hikers to revel in the awesomeness with them. They stop and poke at flowers. They "shhh" the others when they hear a chipmunk. They love the outdoors and want their hikers to love it too.

You must do the same. As soon as you take the client's check, the first part of the hike is over. And that's just not fun.

So after you've established a working budget with your prospect you announce you need some time to think through the system. You want to call the cabinet maker and get his input. You want to check with your staff so they're comfortable installing lighting control into a single zone. And you want some time to think of a few clever ideas you can add to the system to enhance its functionality, and save a few dollars.

Tell the prospect you'll call back tomorrow with more questions.

Now Be Creative.

Certainly there was something the clients mentioned you

had forgotten to take into consideration. Perhaps you noticed something in their house that triggered a creative idea. Whatever it is, dig deeper in to the project, and find a few unique and clever twists to add. Call your subcontractors and ask them for input.

Even better, find ways to bring the cost of individual components down while increasing their performance level. Note: I'm not suggesting you find cheaper components, nor do I favor discounting below-market pricing. I'm suggesting you find clever ways of using higher-value components or processes in the system.

The Callback.

After a day has gone by (and I mean literally, one day) call the prospect and bounce one or two ideas off of him. Like:

> *"You know, we never addressed the master bath. What if we added speakers and a small display in the bath so in the morning you can watch CNN as you're getting ready? Because it's tied into the main system, it'll only add another $3k or so."*

Or you can say:

> *"I spoke with my cabinet maker and he had a great idea. He can build the cabinet to house the front speakers on the lift, so we don't have to put them into the wall. That'll save us installation cost. But it means we'll have to make the cabinet about five inches taller. What do you think about that?"*

These phone questions are your way of double-checking your concepts and keeping your prospect in the loop. As you gain information (i.e., the taller cabinet) you share it with the client. This reinforces your position as a trusted advisor.

When the phone call is coming to an end, ask for another face-to-face meeting. The purpose is to review the system, establish a timeframe for completion, and review the ground rules for working together. It is at this second meeting you finally make the sale.

CHAPTER 9

Selling the Documentation Package.

You've had your initial meeting with the prospect. You've had your follow-up phone call. And now you're having your third encounter. By this time, your prospect is ready to get started, so you need to set the rules of engagement.

Why is the prospect ready to sign? Because he has full information. You told him what you would provide with the Performance List. You proved your capabilities with your Secrets List, you proved you abide by market parity pricing, you told him how much the system costs by Ballparking and establishing a budget of $80k. And you've proven your enthusiasm for the project by calling back a day after meeting them with a few new ideas.

If the prospect doesn't grant you this third encounter, then you've lost the job. Move on.

But if you've done your preliminary work right, he's eagerly anticipating your arrival. And this is what you do: sell him a Documentation Package. And that's all.

Your Only Sale.

Here's how it works. You show up for the meeting and review the Performance List, plus any changes from the past day. In our example, the client loved the idea of the bathroom video system for $3k. You share information you've gathered from your subcontractors and internal staff. You reiterate the new budget, and you tell the prospect about your business practices, including:

1. Billing
2. Site management

3. Follow-through
4. Service

These elements will be discussed shortly.

Do not offer a proposal. Do not hand over a price list, a parts list, or any list for that matter. All you do is talk about the system and establish a budget of $83k. Since the client welcomed you back, he's ready to buy.

One key component of *Companion Selling* is the billing and cash flow strategy, which we'll discuss in ***Chapter 13: Milestoning***. A fundamental precept is a steady trickle of payments – not the unwieldy 30-60-10 payment plans most *Companion Selling* companies use (and abuse). So the first trickle of cash comes from a Documentation Package, which you offer for 5-7% of the established budget.

You tell the prospect:

> *"Our first step is to document and engineer the system. That requires our project manager, Bill, to fully research the system, create electrical and wiring diagrams, put together a parts list with prices, estimate all the labor, and deliver it to us within the next 10 days. Typically, the Documentation Package costs between 5% to 7% of the estimated job, which in our case is, about $4,500. So once I hand that check into my accounting department, the process begins."*

No Proposals. No Estimating. No Spec Work.

You've made the process easy for your client. He was thinking he had to liquidate his money market account to get started, but you only asked for $4,500. That's easy.

Traditional Selling	*Companion Selling*
Selling starts early.	Relationship building starts early.
The proposal is the selling document	No proposals, no spec estimates.

The Documentation Package includes everything you need to build the system. Contrast this to a traditional selling method, where the estimate – often a line listing of products – is used as the selling tool. That's just plain ignorant. Why would you give away your engineering concepts? Why would you spend hours researching prices on automated lifts? Why do any work at all for a prospect? Only clients get work. And they pay for it. Prospects get to talk to you. And that's it.

Prospects get stories. Clients get documents.

Dealing with Pricing.

During your initial conversations with the prospect, you brought up market parity pricing for plasma TV's. Now's your opportunity to explain how your company prices the products you'll be providing.

I recommend you use the term "market price" when explaining your pricing strategy. Plasmas at "market price" are at parity with Best Buy or Costco, assuming the sets are the same. Automation, lighting control, racks, high-end electronics, and speakers at "market price" are typically at MSRP. (I recommend staying away from products easily shopped on the Internet, or available through distribution companies.)

The market price is the price your client would see should they dig on the internet and find a reputable reseller. This is one reason to avoid commoditized products.

Look online for high-end carpet prices; they're almost impossible to find. Now look for furniture from a high-end European design house like Roche Bobois or Ligne Roset. Look up stucco application pricing. Try to find high-end Moen faucets. (You can, but they're priced at MSRP.) These are the elements Chantel sells. She shies away from sofas at Target.com

Honoring the Budget, Part 2.

Your Documentation Package includes a line listing – with pricing – of every component you need to complete the system. Let your client know that when the Documentation Package is delivered, in 10 days, you should schedule a meeting to then review the items and their pricing. This gives your client the opportunity to make product changes on a micro level. But not before then. When pricing questions arise, you can revisit the Performance List and upgrade or reduce elements.

Earlier we discussed the opportunity for your client to engage in pricing discussions. The time to do that is after the Documentation Package has been presented. Talking about individual product prices and labor charges before the Documentation Package is a grave error.

Traditional Selling	*Companion Selling*
Pricing discussions are part of the selling process.	Pricing discussions are held after the Documentation Package is delivered.

Syncing Documentation Package Pricing with Your Ballpark.

If you've done your Backbone spreadsheets carefully, and your Ballparking is accurate, there's a good chance your engineering department will establish a Documentation Package within 10% of your Ballpark.

If the Documentation Package is less than your Ballpark:

Lucky you. Take this information and let your client know how your engineering department found creative ways to reduce the overall project cost.

If the Documentation Package is within 10% of your Ballpark:

Tell your client at the documentation review meeting that

your engineering department found some issues. Offer the client an opportunity to revisit the Performance List, and trim the budget, should they choose. Chances are, your client will accept the cost difference – and move forward with no changes.

If the Documentation Package is more than 10% greater than your Ballpark:

Tell your engineering department to alert you immediately if your Ballpark is short by more than 10%. As soon as you know this information, call your client and let him know what you found. Schedule a meeting to review the Performance List, and see where the errors lie in your Ballparking.

Be prepared to hand back your client's check. If your Ballparking skills are poor, it's fair to say you've misled the client. If you've built a strong enough relationship, and the errors you made are justifiable, then let your client decide to either move forward with the higher budget, reduce the functionality of the system, or back out of the project altogether.

Rules of Engagement.

Companion Selling is a multi-part plan that includes a sales technique and a financial strategy. Using the *Companion Selling* sales techniques without the financial strategy might be fun, but it won't increase your company's profitability. The selling techniques outlined in this book support the financial strategy. You can't have one without the other.

The financial strategy consists of three parts:

1. Developing a modified T&M (time and materials) process.
2. Billing bi-weekly.
3. Treating the sale as a long-term relationship, and not a discrete event.

During your follow-up meeting with the client, you'll have an opportunity to lay out the financial obligations of the relationship, starting with your billing process. We'll discuss these in subsequent chapters.

CHAPTER 10

Modified T&M.

Think back to Chantel. She would never present an estimate for designing a client's house that includes line items and prices. She Ballparks the project, gets buy-in, and starts work.

Her first invoice may be for a shopping trip. Or a swatch proposal. Or a meeting with an architect. It's not for 30% of the estimated project. Chantel operates on a modified T&M (Time and Material) billing process, and so should you.

Risk And Control: The Dentist And the Lawyer.

It's that time again. You reluctantly go to the dentist to get a routine teeth cleaning. You call the office, get a price – let's say $120 – make an appointment, and have the work done. Your dentist doesn't look in your mouth, determine your flossing habits are poor, and jack the price up because it takes another 15 minutes to scrape off a year of mocha lattes.

Your dentist is following a pricing strategy called Fixed Pricing. She's determined that an average person, having a routine cleaning, takes X amount of time, averaging $120 per visit. Some patients, like you, take longer. Some patients – like those who floss – may take less time. Fixed Pricing pushes the burden of risk to the dentist. You risk nothing.

Say you call your corporate attorney to get some legal advice on structuring a real estate transaction. You want to know the tax implications of forming a trust to buy a piece of property. Your attorney prepares a series of documents and sends them to your office for approval. You get an invoice at the end of the month. On that invoice are charges for partner time, assistant time, document preparation, copying and postage.

The attorney tracks his minutes and charges you for each

and every one, plus a few for good measure. In fact, he charges for everything he does for you. I once had an attorney invoice ten minutes for a phone call he made to my office to alert me his secretary had just sent me a fax. The nerve.

Attorneys operate on a T&M billing structure. They have assumed none of the risk in the relationship. Rather, they've shifted the risk entirely to you (which may explain those lawyer jokes).

Choosing how you bill is a balancing act. Traditionally, most installation companies choose the Fixed Price method, like a dentist. They prepare an estimate, have it signed off on by the client, and barring any change orders, hope they're able to make enough money at the end of the project.

That's a losing strategy.

Capturing Every Minute.

It's one thing to offer a Fixed Price on a project that'll take a 90 minutes to complete, it's quite another to offer a fixed price on a project extending over months. If cleaning your teeth were a six month process of meetings, labor, and callbacks, your dentist would switch from Fixed Pricing to something else.

But Fixed Pricing plagues the CI business. It's a hold-over from retail days when every item had a discrete selling price.

I'm sure you've entered labor hours into your proposal database. You figure it takes 1.5 hours to install a keypad. That's nice. And you might even have a Project Management line on your proposal, an arbitrary number derived from some percentage of revenue. As proud as you are of these numbers, I can assure you that at the end of the project, during reconciliation, you're not even close to the actuals. Because installing a keypad includes a host of elements you aren't considering – like setting a meeting with the client to choose the right location. And travel time for the installers. And supervision time to make sure the electrician didn't cut your wires.

Proposal database tools like d-tools are flawed in their treatment of labor because they simplify data entry for "doing" time and make "prepping" time almost impossible to account for.

It takes me ten minutes to stir fry a batch of beef and broccoli. But does it? I need to shop for the ingredients. Since I'm a fussy chef, I insist on getting my beef from a particular butcher. I also like to use CalRose rice, and particular brands of oyster and dark soy sauce. So I have to buy those items from an Asian specialty store 30 minutes from my house.

My Japanese rice cooker takes 20 minutes to make two cups of rice. It takes me five minutes to clean and chop my green onions, garlic and broccoli and 30 minutes to slice and marinate the beef prior to cooking.

I've got to set the table, of course. With plates that had been cleaned earlier. And after dinner? It takes me a good 30 minutes to scrub up. So how long does it take to make a batch of beef and broccoli? Adding in all travel, prep, cook, and clean times: about three total man-hours. And to think I would have entered "10 minutes" into my d-tools database.

To capture all time associated with my dinner, I have to assume I own none of the ingredients, and I'm starting with a clean pantry. Your ingredients are inventory items. My "inventory items" are beef, broccoli, green onions. Your consumables are cable, interconnects, wall plates. My consumables are my sauces (chili garlic, oyster, dark soy), rice wine, sesame and canola oils, water and corn starch.

Your labor includes specifying, purchasing, and receiving your inventory and consumables. I have to pick my ingredients – and shop for them. You have to assemble your inventory items in preparation of the project – the staging and delivery process. I have to stage and assemble my ingredients as well. You have pre-wire and prep work on the jobsite. I have to prep my ingredients by cleaning, cutting, chopping, and marinating. You install, I cook. You program and train, I plate and serve. We both clean up after the work is done.

When broken down to its constituent parts, you see how a ten-minute stir-fry turns into a multi-hour affair. The same is true for your one hour wall plate install. It's more than one hour; you just aren't accounting for all the prep and clean-up time.

These are the time constituents assigned to install labor:

1. Assemble inventory and consumables (Staging & Delivery)
2. Prepare inventory and consumables (Prep & Pre-Wire)
3. Installation
4. Programming & Training
5. Clean Up

Account for these functions when determining your Backbone spreadsheets.

You Can't Predict the Future.

How are you to know if the electrician cuts your wire? How can you tell if the construction schedule slips a few months, requiring extra site visits? How can you predict there won't be changes in product, pricing? You can't. Like the dentist, you try to establish a fixed price by using historical data and padding your numbers just in case.

You figure a project of this scope, with this many components, should take X amount of time. But it's just a guess.

The dentist has more information than you do. She knows you won't have 50 teeth. Because it's a routine, annual visit, she know you don't have seven years of tartar. The tolerance for error for the dentist is +/- 15 minutes. For you, it could be +/- a year.

Estimating High to Cover Risk.

Typical AV companies estimate their labor quite high – typically $75/hour for basic installation, $100/hour for trim, and $150/hour for programming. These prices are several times the actual cost of the labor, but are set high to accommodate the unexpected – and to compensate for inefficiencies.

But these rates simply aren't high enough to earn a good profit on labor. Look at your books, and chances are you're earning less than 20 points of margin on your direct labor. (See ***Appendices A & B*** for a typical breakdown of labor sales, costs, and margins.) In fact, most CI companies break even on labor.

So why not raise your labor rates commensurately? Because if you present an estimate to a prospect with $200 labor rates, you'll be laughed at.

Develop Truer Labor Costs.

Companion Selling relies on a modified T&M billing process, allowing you to charge your client for every labor hour expended at a fair rate. I recommend you set your labor rates to achieve a 50-point gross margin, regardless of efficiencies.

So if you're paying your lead tech $30/hour, and his annual benefits and project-related T&E costs average $8/hour, than your cost is $38/hour, and you should bill him out at $76/hour. That'll give you 50 points of margin. See ***Appendices A & B*** for a complete discussion on determining labor costs.

The same math holds true for any of your direct labor employees. Now you've got a baseline of labor rates.

Traditional Selling	***Companion Selling***
Labor rates are set to "market" and include a risk factor.	Labor rates are set at 2X loaded cost.
Actual labor margins differ from proposed. Margins are impossible to predict.	Labor margins are 50 points.

Use Your Low Labor Rates As a Selling Tool.

Explain to your client how you establish your labor rates, and why they are significantly less than the competition. Tell your client how the industry, in its inability to predict the future, jacks up its rates as protection. In comparison, your company uses a unique billing process to keep labor costs in check, reduce risk, and ultimately provide a higher value project. That billing process is outlined in the next chapter.

Revisit Chapter 6: Building the Backbone.

Refer to your Backbone spreadsheets. Plug in your labor hours, accounting for the five constituents outlined above. Your hours will increase substantially, but the selling price per hour will go down. The net effect should be an increase in labor per project. This net gain is the profitability key to *Companion Selling*. You must front-load these labor costs into your Ballpark estimates in order to capture them later when you bill.

Will your Ballpark estimate be higher than a competitor's? Perhaps. But only you have done the relationship-building work, built a Performance List, and given the prospect an opportunity to interact with the budget. Your Ballpark is no longer the most critical element of the sale. It's just an indication of value.

Companion Selling reduces the importance of price in the selling process, and it redirects the prospect to the important parts of the project – design, functionality, performance, and fun. Price, as you know, is a fluid part of the sale.

Go back to our example of selling a system to your mom. Are you concerned about price when you make the presentation to your mom? Of course not. You're thinking about ease of use. You're thinking about designing a system so robust you won't get a service call in the middle of the night. You're thinking of optimizing your mom's AV experience. Price is secondary.

The same is true in Companion Selling. You begin to consider price as a secondary element in the process.

CHAPTER 11

Invoicing.

Tell your client your company will present an invoice every two weeks. In the invoice will be a description and price for the labor used for the previous 14 days. Also on the invoice will be the cost of any materials you need to purchase. It's that simple. Each invoice looks like the example on the following page.

Why is this a modified T&M billing system? Because you are working within the budget. The budget that you Ballparked and verified with your Documentation Package. Like the dentist, there's a cap on your invoices. But like the attorney, you have some wiggle room.

Companion Selling is an enormous profit generator if done properly. It allows you to capture every labor hour you expend on a project. Right now, you're losing most of those labor hours by throwing them into a catchall category called "management."

But look closely at the AV Solutions invoice. Every time Tom the project manager meets with the GC or a subcontractor, he notes that time, describes the nature of the meeting, and charges for it. Sure, some of those meetings are $37.50, but that could be $37.50 you didn't collect previously.

Everything Gets Charged. Everything.

The labor invoice includes every hour your company spends on the project, including site visits, shopping trips with the client, meetings with the general contractor, time spent waiting for the cable guy, and of course, install, trim, programming, supervision, training, and follow-up time.

Every hour is charged. Every hour is justified. Notice how labor is charged on the AV Solutions invoice. Techs are at $58/hour. The project manager is $75/hour. These rates are 2x

AV Solutions
742 Evergreen Terrace
Springfield, CA 92121

Client Thornton Melon
Project Bedroom Surround System

Labor Completed For The Past 2 Weeks

Date	Work Performed	By	Hours	Rate	Ext.
May 12	Client Request: relocate entryway keypad to left side of door.	BP Tech	1.5	$58.00	$87.00
May 12	Complete wiring of master bath speaker system.	BP Tech	3	$58.00	$174.00
May 15	Site visit with GC. Discussed electrical conduit placement, drywall schedule.	TL PM	0.5	$75.00	$37.50
May 17	Conference with electrical sub. Discussed placement of conduit for MB wiring.	TL PM	0.5	$75.00	$37.50
May 19	Meeting with cabinet shop. Presented dimensioned drawings for plasma, lift, speaker system. Found new plasma lift vendor to customize rack and pinion system. Will save $200 vs. estimate.	TL PM	1	$75.00	$75.00
May 22	Completed wiring master bedroom.	BP, JD Tech	4	$58.00	$232.00
Total Labor					$643.00

Product Required For Next Stage

Qty	Description	Price	Delivery	Tax	Ext.
1	Bay Audio LCR Bar for 60″ plasma. Custom finish to match.	$2,000.00	$25.00	$150.00	$2,175.00
2	Bay Audio JM65 speakers for surround.	$675.00	$10.00	$50.63	$1,471.25
1	Middle Atlantic rack system for bedroom closet.	$825.00	$26.00	$61.88	$912.88
Total Product Required For Next Stage					$4,559.13
Invoice Total					**$5,202.13**

each employee's pay, including benefits. That's 50 points of margin.

Now look at the description of their work. The techs noted a client request to move a keypad. The PM noted a site visit with the GC. These are the little things that suck up your labor. But they are a necessary, and frequent part of the project. That's why your client should pay for these services.

We'll look at client requests in ***Chapter 12:** Change Orders*, and non-standard charges in ***Chapter 13:** Milestoning*.

Track Hours Internally.

Every employee associated with the project must maintain a time log – like an attorney – with a detailed description of their activities.

Be sure to capture all time spent on the project. When you do, you'll find your labor margins increasing substantially as your utilization and efficiency increases. Use the time log internally to verify presuppositions about installation times – and use the newly acquired information to improve your Ballparking skills.

Expect to See Profits and Cash Flow Increase Dramatically.

By using the *Companion Selling* financial tools, you will see immediate changes in your financial performance. As compared with traditional selling and billing systems, *Companion Selling* will increase your labor efficiency by pointing out – through meticulous time tracking – how many billable hours you have per day.

Inefficiencies and overcapacity call attention to themselves within a month's time, so measure your labor utilization on a monthly basis, by the using the formulas in **Appendices A & B**.

Explain Your Billing System.

Explain how your invoicing system helps keep track of labor and smooths out the client's cash flow requirements. They'll like that.

By invoicing every two weeks, you'll keep the value of each

invoice lower, making it easier for the client to make payment. Let's return to our bedroom surround system at $80,000. A typical 30-30-30-10 billing plan requires the client to write three $24k checks and one $8k check.

These are big chunks and cause consternation with some prospects. First, why are they writing a $24k check to get started? It seems like a huge commitment. And second, they believe they need $80k sitting around before they get started with you. Your revenue stream for a six-month project might look like this:

	Month 1	Month 2	Month 3	Month 4	Month 5	Month 6
Revenue	$24,000		$24,000	$24,000		$8,000
Stage	Pre-wire		Product	Product	Trim	Final
Cash	High	Medium	Medium	Medium	Medium	Low

Should you bill every two weeks, you might see an invoice stream like this:

	Week 1	Week 3	Week 5	Week 7	Week 9	Week 11	Week 13
Revenue	$5,600	$3,000	$3,000	$3,000	$20,000	$10,000	$10,000
Stage	Eng.	Pre wire	Pre wire		Product	Product	Product
Cash	Medium	Medium	Medium	Medium	Medium	Medium	Medium

	Week 15	Week 17	Week 19	Week 21	Week 23	Week 25
Revenue	$8,000	$5,000	$5,000	$3,000	$2,500	$1,900
Stage	Trim	Trim	Trim	Trim	Final	Final
Cash	Medium	Medium	Medium	Medium	Medium	Medium

You can explain how your billing schedule reduces the "big shock" checks. They'll appreciate that. In fact, they might be able to write many of the checks from their own cash flow – so you'll reduce the trauma of those big payments.

In addition, your clients retain more control of the process. By only paying for completed labor and product required for purchase, they've reduced their exposure. This is a comforting feeling.

Bi-Weekly Billing Allows the Client to Buy More.

When you break your invoicing into manageable chunks, the client understands how each invoice is associated with the work you've done. You install something, they get charged. You make a site visit, they get charged. They make a request, they get charged. It's all apparent.

So as the process goes along, the client is more apt to make requests, increase the scope of the project, and demand higher-performance components. Because invoicing occurs incrementally, it's really never "too late" to make changes. Compare this to a 30-30-30-10 plan, where each payment locks the client in. Changes aren't as easy to make.

Traditional Selling	*Companion Selling*
Invoicing occurs in arbitrary amounts at preset times.	Invoicing tracks labor expended, products required.
Invoicing discourages upgrades and improvements.	Invoicing encourages upgrades and improvements.

One of the wonderful aspects of the AV industry is its flexibility. Because the industry grew up accommodating new construction and retrofit, products and processes are appropriate for both. That means changes are easy to make at virtually any stage of completion.

Compare this to plumbing. The plumber wants all his pipe work done before the drywall is installed. The flooring guy has one opportunity to install tile in the bathroom. The cabinetmaker gets one shot, too. But the AV Integrator can make modifications to the system from pre-construction through final painting.

Having that flexibility allows you and your client to upgrade, change, and improve the system as the home is being built. By using modified T&M billing, you reduce the number of cut-off dates. The installation becomes more fluid, and you can react faster to changing desires as they occur.

CHAPTER 12

Change Orders.

Change orders are an odd way of punishing the client for being enthusiastic. "Sure, we can move the keypad to the other side of the doorway, but that's a change order, and we'll have to charge you for it."

Those are ugly words, and certainly they violate every rule of *Companion Selling*. There's no inclusive language. It draws a line between the client and your company. It's just wrong.

Let's not think of these as changes. Let's think of them as client requests, part of the natural process of creating the perfect system. Since you're invoicing every two weeks – and you're filling each invoice with a detailed explanation of your labor, it's appropriate to write about client requests as they occur.

For example, the invoice line might indicate this:

Date	Work Performed	By	Hours	Rate	Ext.
May 12	Client Request: relocate entryway keypad to left side of door.	BP Tech	1.5	$58.00	$87.00

See how much nicer that is? No antagonism. No finger pointing. And since you invoice every two weeks, these changes – or rather, "client requests" – are handled when they occur – and not at the end of the job when they're likely to be questioned.

Recently I had construction work done in my office. The GC gave me a rough estimate of the cost of the work. As the work progressed, I thought of additional improvements. I discussed the improvements with the GC and he sent me a confirming email noting the incremental dollar figure associated with the work.

He didn't remind me the changes would put me over budget. Obviously, I'm aware my changes upped the budget. I don't need to be scolded. He happily completed the work and charged the incremental fee. That's how you handle a client request.

CHAPTER 13

Milestoning.

It's your responsibility to keep tabs on the progress of the project and track costs against the completion schedule. If you see the project slipping and costs creeping up, you have to find out why and inform your client. Immediately.

Remember: you are your client's advocate. You are his CFO. You are looking out for his best interest at all times.

This does not mean you begin discounting if the project starts to go over budget. It means you constantly track the project against your Documentation Package and speak up as soon as a problem arises.

Milestoning at natural break points is a great way to verify the project is moving along as expected. So set milestones after prewire, before sizable product orders, after significant trim stages, and after programming. Make note of client requests as they occur and bring them to your client's attention if they have a material affect on the budget. As discussed in the previous chapter, you'll be noting client requests on your bi-weekly invoice. Make sure your client understands these requests are "above and beyond" the budget outlined in the Documentation Package, but don't keep reminding them if they're "over budget." It's condescending.

Milestoning Catches These Four Problems.

There are four primary reasons why you have a tough time making money on fixed bid projects.

1. Change Orders Go Unnoticed.

Most CI companies catch the big change orders (which we now call client requests), but the little ones get subsumed into the process. Why? Because someone in your company feels guilty. Here's the psychology: Your technician is working on a $100k project. He likes the client. The client asks, "Can you move the wall plate to the left side of the door, please? I didn't realize it was going on the right side."

So the technician has a choice. Penalize the client for a change that'll take 90 minutes, or tell the client, "No problem, I'll take care of it right now." Of course, your tech won't ever tell the accounting department to bill for the time. It's his "gift" to a nice client on a big job.

The client didn't ask for a gift. The client wanted the keypad moved. But the tech did the right thing under the circumstances. He found a way to keep the client happy, because change orders are a form of punishment. If the tech said, "Sure, I can move the keypad, but that's a change order," the client would have been offended, and the relationship you worked so hard to establish is compromised.

There are dozens of these small changes per project, and at 90 minutes each, they represent a lot of unbilled time. Bi-weekly itemized billing catches each of these changes without penalizing the technician or aggravating the client. Your client will see the cost of these incremental changes in a gentle way.

2. Site visits go unbilled.

Long-term projects usually involve lots of other trades. These tradespeople are unfamiliar with the AV business and delight in cutting, mangling, or shorting strange looking wires. To protect your labor investment, you must have a project manager walk the jobsite on a regular basis. These visits protect you, of course, as much as they protect the client. You should account for these site visits in your Backbone spreadsheets and include them in your Ballpark figure. So charge for them.

3. Contractor coordination goes unbilled.

Meetings with the GC, the electrician, the flooring guy, the cable guy – these are all billable. In fixed bid contracts, the number of management meetings built into the quote is a wild guess, and usually a low guess. Bill for every one of these meetings, and be sure to note the nature of the meeting, the attendees, and the outcome on your invoices.

During the Documentation Package, inform the client of the estimated number of contractor meetings included in the budget, but charge for every one as they arise. Make note on your invoice of "additional" contractor meetings beyond your original estimate, and note why they occurred. Often these extra meetings are the result of a change order the client requested through another subcontractor. For example, the client asks the GC to move the location of a patio door. That change alters the placement of an in-wall speaker. Capture this time and invoice for it accordingly.

4. Projects don't stay on schedule.

A 40-hour project completed in one week is twice as profitable as a 40-hour project stretched over four weeks. This is because every visit to the jobsite requires travel time. In addition, every time your techs restart a project, it takes some ramp-up time to reacquaint themselves with the status of the project. Because you can't predict how long a project runs, you'll capture the inefficiencies of a stretched-out project through your bi-weekly billing. How? Because your techs may take an extra 30 minutes reacquainting themselves with the project, and that's captured in their time logs as billable hours.

Think Like the GC.

There's no doubt client requests, additional site visits, contractor coordination, and drawn-out projects add costs. In the world of fixed budgeting, you pay for a lot of these additional costs. But bi-weekly billing and Milestoning let you capture all the time spent on the job, charge for it appropriately, and keep the client informed along the way.

The GC is doing it. The electrician does it. You should, too.

Milestoning keeps everybody
on the same page.

CHAPTER 14

Cash Flow Benefits of Companion Selling.

Cyclical and sporadic cash flow is the bane of the CI business. In this chapter, I will attempt to outline the troubles most CI companies have with cash and show how *Companion Selling* helps smooth these cash flow problems, increase profitability, and put more money into the pockets of the employees.

We'll use the fictitious company AV Solutions for our examples. But first, some definitions.

Financial Term	What It Means
Product Sales	Income only from the sale of products.
Labor Sales	Income from the sale of any labor. Includes engineering, documentation, installation, programming, and service. Also includes labor you buy from subcontractors.
COGS	Cost of Goods Sold Product Cost of Goods Sold. This is the gross cost of products you purchase. Does not include freight costs, rebates or discounts. Labor Cost of Goods Sold. This section includes direct labor payroll, benefits, and project-related time and materials costs.
EBITDA	Earnings Before Interest, Taxes, Depreciation and Amortization. This represents cash flow from operations.

AV Solutions: Smooth Cash Flow

	Month 1	Month 2	Month 3	Month 4	Total
Product Sales	66,667	66,667	66,667	66,667	266,667
Labor Sales	16,667	16,667	16,667	16,667	66,667
Total Sales	**$83,333**	**$83,333**	**$83,333**	**$83,333**	**$333,333**
COGS Product					
COGS	38,667	38,667	38,667	38,667	154,667
Freight In	1,547	1,547	1,547	1,547	6,187
Freight Out	773	773	773	773	3,093
Purchase Disc.	(1,160)	(1,160)	(1,160)	(1,160)	(4,640)
Total COGS Product	**$39,827**	**$39,827**	**$39,827**	**$39,827**	**$159,307**
COGS Labor					
Direct Labor Payroll	12,083	12,083	12,083	12,083	48,333
Direct Labor Benefits	1,208	1,208	1,208	1,208	4,833
Project Related T&E	1,833	1,833	1,833	1,833	7,333
Total COGS Labor	**$15,125**	**$15,125**	**$15,125**	**$15,125**	**$60,500**
Total COGS	**$54,952**	**$54,952**	**$54,952**	**$54,952**	**$219,807**
Gross Margin	**$28,382**	**$28,382**	**$28,382**	**$28,382**	**$113,527**
Expenses					
General & Admin.	20,833	20,833	20,833	20,833	83,332
Sales & Marketing	2,083	2,083	2,083	2,083	8,332
Rent, Utilities, Phone	4,167	4,167	4,167	4,167	16,668
Total Expenses	**$27,083**	**$27,083**	**$27,083**	**$27,083**	**$108,332**
EBITDA	**$1,299**	**$1,299**	**$1,299**	**$1,299**	**$5,195**
Accounts Receivable	69,444	69,444	69,444	69,444	
Accounts Payable	37,507	37,507	37,507	37,507	
Cash Used	(1,299)	(1,299)	(1,299)	(1,299)	
Cash Balance	**$1,299**	**$2,597**	**$3,896**	**$5,195**	

Note: Numbers have been rounded for clarity, and may not add up properly.

Meet AV Solutions.

This is a typical AV integrator with approximately $1m in annual revenues and six full-time employees. The owner acts as main salesperson. There's a project manager who spends some time in the field. He also does programming and handles some design and engineering tasks. There's an office manager, a lead tech, and two junior techs. (See ***Appendix C*** for AV Solutions' modified P&L.)

AV Solutions: Cyclical Cash Flow

	Month 1	Month 2	Month 3	Month 4	Total
Product Sales	66,667	46,667	86,667	66,667	266,667
Labor Sales	16,667	11,667	21,667	16,667	66,667
Total Sales	**$83,333**	**$58,333**	**$108,333**	**$83,333**	**$333,333**
COGS Product					
COGS	38,667	27,067	50,267	38,667	154,667
Freight In	1,547	1,083	2,011	1,547	6,187
Freight Out	773	541	1,005	773	3,093
Purchase Disc.	(1,160)	(812)	(1,508)	(1,160)	(4,640)
Total COGS Product	**$39,827**	**$27,879**	**$51,775**	**$39,827**	**$159,307**
COGS Labor					
Direct Labor Payroll	12,083	12,083	12,083	12,083	48,333
Direct Labor Benefits	1,208	1,208	1,208	1,208	4,833
Project Related T&E	1,833	1,833	1,833	1,833	7,333
Total COGS Labor	**$15,125**	**$15,125**	**$15,125**	**$15,125**	**$60,500**
Total COGS	**$54,952**	**$43,004**	**$66,900**	**$54,952**	**$219,807**
Gross Margin	**$28,382**	**$15,330**	**$41,434**	**$28,382**	**$113,527**
Expenses					
General & Admin.	20,833	20,833	20,833	20,833	83,332
Sales & Marketing	2,083	2,083	2,083	2,083	8,332
Rent, Utilities, Phone	4,167	4,167	4,167	4,167	16,668
Total Expenses	**$27,083**	**$27,083**	**$27,083**	**$27,083**	**$108,332**
EBITDA	**$1,299**	**($11,753)**	**$14,351**	**$1,299**	**$5,195**
Accounts Receivable	69,444	59,028	69,444	79,861	
Accounts Payable	37,507	26,255	48,759	37,507	
Cash Used	(1,299)	12,589	(26,438)	20,370	
Cash Balance	**$1,299**	**($11,290)**	**$15,148**	**($5,222)**	

Note: Numbers have been rounded for clarity, and may not add up properly.

You Want Your Cash Slow and Steady.

Smoothing out your cash receipts often increases the amount of cash on hand. These spreadsheets show how.

The first spreadsheet shows AV Solutions selling the exact same amount each month. It's an unrealistic situation to be sure, but one that proves a point. Note it has $1,299 in the bank at the end of month one and $5,195 in the bank at the end of the fourth month. Its bank balance matches its EBITDA because accounts payable and accounts receivable remain constant month to month.

Now look at AV Solutions with the same dollar value of sales over the four-month period, but each month is either higher or lower than the previous one. This mimics the natural cyclicality of a business.

In this case, AV Solutions has a negative bank account balance at the end of the four months of ($5,222). That's a $10,417 swing from the non-cyclical example.

AV Solutions earned just as much money – EBITDA is still $5,195. Sales in the four-month period are the same at $333,333.

But the cyclicality of the business increased the *use* of cash – note how accounts receivable and accounts payable differ. There's simply less cash available in the bank on a month-to-month basis. That's cash that could have been used for increased salaries, capital expenditures, a stronger marketing program.

Because every business has fixed monthly costs, those months with lower sales put a greater strain on cash reserves. True, the strong sales months fill the coffers back up, but the constant up and down of the cycle means you'll have less cash at your disposal – on average – then when sales are smoothed out.

In the AV Solutions example, you see the difference between smooth and cyclical sales. Both have the same four-month revenue, and both have the same cost structures and expenses, but the smooth scenario has just enough cash to continue with operations. The cyclical scenario needs another $11,290. Look at the cash balance line in the cyclical scenario: this is the negative amount in month two. AV Solutions needs to borrow $11,290 this month in order to make payroll.

AV Solutions: Smooth Cash Flow

	Month 1	Month 2	Month 3	Month 4
Cash Used	(1,299)	(1,299)	(1,299)	(1,299)
Cash Balance	**$1,299**	**$2,597**	**$3,896**	**$5,195**

AV Solutions: Cyclical Cash Flow

	Month 1	Month 2	Month 3	Month 4
Cash Used	(1,299)	12,589	(26,438)	20,370
Cash Balance	**$1,299**	**($11,290)**	**$15,148**	**($5,222)**

Companion Selling, along with the two-week billing cycle, helps smooth your revenue stream. A smoother revenue stream reduces the peak demand for labor. You may find you can reduce your labor force – or better yet – grow without adding labor – because a smoother workflow allows you to fine-tune your capacity.

Profit is Not Cash.

There's a misconception net profit is the same as cash. It's not, for several reasons: working capital (consisting of accounts receivable, inventory, and accounts payable), interest payments (to the bank and investors), taxes, and capital expenditures.

If your company was paid cash with every invoice, like a restaurant, you'd have no accounts receivable. If you, in turn paid for products COD when the UPS truck arrived, you'd have no accounts payable. And if you only bought products for a specific project, you'd have no inventory. With these three accounts at zero, net profit before capital expenditures does indeed equal cash.

But most CI businesses have accounts receivable. Often a month's worth of receivables are awaiting collection. If you sell $100k in one month, you earn 10% EBITDA, but it takes you 30 days to collect your money. On average, your numbers look like this:

	Month
Total Sales	$100,000
EBITDA	$10,000
Accounts Receivable	$100,000
Cash Balance	**$0**

You have $100k in receivables, a $10k profit, but no cash in the bank generated for the month.

If you owed your vendors $10k for the month but decided not to pay, you now have an extra $10k in the bank. That's why cash is reduced as receivables grow, and cash is increased as payables grow. The opposite holds true. As your clients pay sooner, your receivables shrink and your cash grows. But as you pay your vendors sooner, your cash shrinks.

Since most CI companies don't maintain inventory, we'll leave the cash affect of inventory out of our discussions. In short, as inventory increases, cash decreases.

But CI companies do buy computers, testing equipment, and vehicles. These are capital expenditures, and are a use of cash. Capital expenditures, like buying a panel van with cash, show up on your balance sheet as assets. As you depreciate this asset (that is, use it up), you expense the depreciation amount each year. That expense is a non-cash event (remember, you bought the truck already). Depreciation is a tax event, which is why we don't count it in our cash flow strategies. Amortization is a corollary to depreciation – it's depreciation for non- physical purchases like intellectual property, patents, and start-up costs.

Now you know why we look at EBITDA after adjusting for accounts receivable and payable. That's cash.

Profit is Good. Cash is Better.

A healthy profit indicates your business is fundamentally strong. But a healthy checking account balance in addition to a profit means your company is fiscally sound. Cash in the bank gives you the flexibility to respond to opportunities and weather a bad cycle or two. *Companion Selling* increases cash flow by smoothing out income, matching variable expenses to income, and reducing accounts receivable.

CHAPTER 15

You Are the Client's CFO.

You've collected your Documentation Package check. You've had your engineering staff complete its wiring diagrams and product list. You've had your review meeting, and the project is underway. Now's not the time to revert back to the old ways of a contractor.

Unlike a specifier, Chantel sees each project as a small part of a greater whole. The greater whole, of course, is a lifetime of designing with the client. Today's project is a living room. Tomorrow's project may be the den. And next year? The summer house.

Chantel develops relationships with clients. She doesn't sell a "designed living room."

Companion Selling suggests you develop relationships as well and stop thinking of yourself as "the guy who sells home theaters." Incumbent upon you is to keep your enthusiasm and knowledge high for audio, video, and automation solutions. Allow yourself to consider and suggest alternatives, options, modifications, and improvements to any system on which you are currently working. Broach the subject of AV and automation for your client's other homes. See yourself working hand-in-hand with this client far into the future, and let that concept drive you to engage them in discussions of automation, interface, and performance.

Spend Your Client's Money As If It Were Your Own.

With your relationship comes great responsibility. As a trusted consultant, you must provide the highest-value designs for your client. That means digging deep into the tricks of the trade and fully sharing your knowledge. As Chief Financial Officer, you are ultimately responsible for every dollar spent. Make sure you're spending wisely.

Are you aware of a high-value plasma TV? Then use it. Do you know how to save labor by having the GC handle some of the work? Then take advantage of this. Each opportunity you have to save the client money – while providing a superior solution – must be taken.

Spend your client's money as if it were your own. Make good choices. Don't experiment on their dime. And be sure to note discounts and savings on your bi-weekly invoices, like this:

Date	Work Performed	By	Hours	Rate	Ext.
May 19	Meeting with cabinet shop. Presented dimensioned drawings for plasma, lift, speaker system. Found new plasma lift vendor to customize rack and pinion system. Will save $200 vs. estimate.	TL PM	1	$75.00	$75.00

Spend Wisely: Buy the Best.

Do not confuse value with price. I'm not suggesting you purchase the cheapest plasma TV from your local discount warehouse store. In fact, I suggest you find the best quality plasma TV you can. But make sure you're getting the best value pricing on the highest performance items and passing those savings on.

Chantel doesn't think to shop at Pier 1 for furniture. She leads her clients to designer showrooms, with exotic fabrics, imported pieces, and customized finishes. These are exciting. You should lead your clients to exotic, exciting, high performance solutions as well. Get the best stuff you can, for the best prices.

Chantel keeps her enthusiasm high. She gets her clients enthused. What causes excitement? Items out of the ordinary. Items that are novel and unique. Items that perform above the field.

Look in your client's home. There's a Viking range, perhaps.

And a Nautilus washing machine along with tumbled travertine flooring. Look at the pull knobs on their kitchen cabinets. Their bathroom fixtures. Most likely, none of these are Home Depot parts. They're a step up. Unique. Novel.

When your clients are building their homes – or designing their spaces, they face a multitude of decisions, including flooring, tiling, appliances, wall treatments, moldings, lighting fixtures. Rest assured that your high-end clients take every one of the decisions seriously. They typically choose solutions above the ordinary.

This makes them feel good, of course. It's a way of surrounding themselves with items above the fray and reaffirming their own position in society – above the fray.

This is your entrée into unique and novel AV solutions. As a rule, avoid any products available in a mass retailer, through national distribution, or available on the internet. These are not unique products – these are the ho-hum everyday products that don't create excitement.

Buy unique, high-performance, well-built products. Avoid the common.

Share Your Knowledge.

Every time you find one of these high-performance values, share your excitement with your client. First off, they love to hear about your conquests. They appreciate how you conserve their funds. And you've given them a great story to repeat to friends:

> *"So my AV guy found this 100" plasma TV – it's the first one on the market – and was able to negotiate a discount from the manufacturer in exchange for having our home photographed for their web site. Isn't that great?"*

That's how a CFO operates.

Spend your client's money
as if it were your own.

CHAPTER 16

The Unwilling Client.

What happens if your prospect doesn't respond to your *Companion Selling* techniques? Let's say she isn't interested in going on your hike. She just wants a proposal, plain and simple.

You've got a few choices. You can return to traditional selling, estimating, and billing approaches, but you risk losing the bid and your ability to manage expenses and revenue through the modified T&M billing process.

Or, you can condense the *Companion Selling* relationship-building aspects and move straight to the Performance List. Building a Performance List is required in any selling approach. The subtlety with Companion Selling is how you ease into the process. Since you've built a working relationship with your client first, building the Performance List is a natural give-and-take process your client will welcome. If you forego the relationship-building, then building a Performance List may sound a lot more like order taking.

Next, you should Ballpark and Honor the Budget. Even the most unwilling client has a budget in mind and is concerned you aren't sensitive to it. From there it's smooth sailing to your Documentation Package and setting forth your modified T&M billing process.

These four elements ensure your sustained profitability. The relationship-building aspects of *Companion Selling* are important. But there are prospects who are wary. So in those cases, reduce the amount of time you spend developing the relationship – and move on to the Performance List. But never forego the financial aspects of *Companion Selling*. You'll quickly return to low profitability if you do.

So, if your prospect eschews the relationship-building

aspects, but agrees to develop a Performance List, set a budget and pay for the Documentation Package. Then it's full steam ahead.

If the client won't participate in developing a Performance List, setting a budget, and buying the Documentation Package, then decline to do the work. The prospect is telling you up front you'll be accepting all of the project risk. You'll also be denied your due profit.

An unwilling client is not interested in relationship-building. Keep all other elements in place, and proceed.

When the Client isn't Accessible.

Often on larger projects you do not meet the client: you work directly with the builder, the architect, or an owner's rep. Great. These go-betweens are your client's proxy. Treat them as you would a prospect, and use your *Companion Selling* techniques, modified of course, for their position.

> *"I understand we'll be working together on this bedroom surround system. It should be a fun project. Let's talk about what the homeowner would like to achieve in this room."*

The presentation doesn't change. You just have to reference the homeowner continuously. When you ask Performance List questions, expect the owner's rep to say, "I don't know, but I'll get back to you on that."

An enthusiastic owner's rep is your best salesperson, so don't skimp on the relationship. If she's sold on a concept, she'll sell it to the homeowner.

Having an owner's rep in the loop doesn't change the strategic elements of *Companion Selling*. Stick to the plan and utilize the financial rules. These rules are your guarantee of profitability and should become a permanent fixture of your company.

When the Project is Quick.

Some projects turn around so quickly – even big dollar projects – that the *Companion Selling* process becomes cumbersome. What part of the process should you cut?

As a review, there are five important stages in Companion Selling:

1. Establish a relationship.
2. Develop the Performance List, Ballpark, and test against the budget.
3. Sell the Documentation Package.
4. Begin work and bill bi-weekly.
5. Milestone to keep on track.

Even when a project has a quick turn-around, you still have to establish the relationship with the prospect. You have to develop a Performance List. You have to Ballpark and compare to budget.

You can choose to forego the Documentation Package. Here's where you can save time, because your first invoice may include product you need right away. Thereafter, billing every two weeks won't slow things down, so it's best to continue that practice to keep cash flow smooth.

Quick turn-around projects have all the elements of long drawn-out projects, they're just condensed in time. Condensing the billing process makes sense, but there's no reason to eliminate the relationship-building tools. These will serve you well in the future when the client has a new large-scale project. They'll turn to you because you have a relationship.

When the Project is Small.

You're doing small projects – why? If it's a favor for an existing client, you already have the relationship, so just Ballpark – and start your Modified T&M billing. If it's a new client who has the potential to be a big, then introduce your working style now with the complete *Companion Selling* technique.

It's worth investing in client development. Perhaps the first project you do for them is small, or perhaps it's a quick turn-around project. No matter. You have to amortize your relationship-building time against the long term.

If you believe this is a one-time client, then I recommend you forego the project. Why? Because you're building a long-term relationship-oriented business with *Companion Selling*. So why divert your energies to a one-time project?

CHAPTER 17

Companion Selling Summary.

Companion Selling helps forge a bond between the seller and buyer by taking an "arm-in-arm" approach to a project.

Traditional sales techniques rely on persuasive language to move a buyer from one position to another. *Companion Selling* uses inclusive language to undertake the journey together.

The seller takes on the role of trusted trail leader, leading the buyer through the process. Everybody is hiking, but the trail leader knows the best route. And like the trail leader, the *Companion Seller* talks about the excitement of the journey as well as the destination point.

A Brief History of Selling, and Why It Keeps Failing in Our Industry.

You've seen these traditional selling techniques in action:

Traditional Closing Techniques.

These tried-and-true selling tools seek ways to justify the purchase in the client's mind. Read *The Prentice-Hall Miracle Sales Guide,* a classic manual on closing styles.

Consultive Selling.

An update on traditional closing techniques, consultive selling takes a problem-solving approach to a sales situation. The salesperson presents clever, unique, or proprietary solutions – and then uses traditional selling techniques to justify the sale. Think *How to Win Friends & Influence People* by Dale Carnegie.

SPIN Selling and Derivatives.

Popularized in the late '80s, these "psychological" selling

techniques split the sales process into types of customer questioning. These techniques try to have the customer "sell himself" by answering pointed questions that lead down a predetermined path. *SPIN Selling* by Neil Rackham started the trend.

The Trouble with Traditional Selling Techniques.

Traditional selling techniques are stuck in the retail/wholesale world and have no relevance in our industry. These techniques are product based, and rely on language to help assess and then solve a problem. They try to move a product or service from one side of the trading line to the other.

Traditional selling techniques acknowledge the adversarial relationship inherent in the selling process and find ways to reduce the tension between buyer and seller.

Companion Selling Breaks The Mold.

Companion Selling reduces the role of products and increases the role of adventure and accomplishment. It relies on enthusiasm, empathy, and shared goals between seller and the buyer.

Companion Selling redefines the adversary. No longer salesperson v. client, but salesperson & client v. The Big World.

Companion Selling allows the seller to share in the ups and downs of the accomplishment.

Companion Selling Redefines The Selling Process.

The seller takes on the role of "Project CFO," searching for the best value for the client.

The seller and the client both present solutions to each other. The seller checks the solution for value and probability of success. The seller allows the client to have aesthetic and functional input, but steers the input accordingly.

Companion Selling sees the sale as a series of small events:

1. Engineering and Documentation

2. Craftsmanship (carpentry)
3. Product acquisition, programming
4. Installation, tuning, teaching

Companion Selling Vocabulary.

Review this vocabulary used in the *Companion Selling* presentation. These are the kinds of comments you'll be making to your prospective client. Once again, let's use our bedroom surround system as an example:

- *How involved would you like to be in this project? We can do it together, or if you're more comfortable, I can handle it all for you.*
- *This is going to be fun. We're going to have a great time doing this together!*
- *Is your husband/wife going to join us, or are we working alone?*
- *Have you seen any bedroom surround systems you really like?*
- *Let's write down our performance and functionality goals.*
- *What if we had motorized shade and lighting controls? Would you like that?*
- *I know a great carpenter who can build beautiful furniture, and save us money. We can hide a plasma on a motorized lift right in the furniture.*
- *Let's talk about other aesthetic considerations, like speakers and keypads.*
- *What's our budget?*
- *My best guess puts a bedroom surround system with this level of performance at $100,000. I know that's higher than your budget, so let's go down our list together and start paring things back.*

- *We should start by taking out the lighting control. It costs a chunk of money, and I think our budget is better spent elsewhere.*
- *Now that the Performance List and budget are in agreement, we need to take the next step, which is a fully engineered design of the system. This includes a wiring diagram, a complete parts list, and a set of budget guidelines for each major step of the way. This documentation and engineering process requires a $7,500 deposit to get going. When can we start?*

Step by Step Through the *Companion Selling* Process.

1. Use inclusive language when discussing projects with your prospect.
2. Offer secrets from your Secrets List.
3. Make sure the client values the secrets and begins to use inclusive language with you.
4. Write a Performance List together, including:
   - Performance criteria
   - Functionality
   - Aesthetic concerns
5. Use "What If" questions to introduce new technologies and concepts.
6. Ask about the budget.
7. Ballpark the project, compare to the budget, and begin to pare down the Performance List, assuming the budget is a reasonable number.
8. Describe market pricing for products, and how your labor rates are established.
9. Allow the client to re-evaluate by reducing the scope of the Performance List, or increasing the budget.

10. When the Performance List and budget are in agreement, excuse yourself from the meeting and promise a call-back in 24 hours.
11. Call back the next day with additional questions, comments, and secrets. Set a follow-up meeting to discuss the Performance List and budget one last time.
12. Re-present the Performance List and confirm the budget.
13. Re-present market pricing, labor pricing. Present bi-weekly billing.
14. Sell the Document Package.
15. Turn the project over to a project manager.
16. Begin billing on a bi-weekly basis.
17. Milestone as the project progresses to keep projected and actual costs in line.

Things a *Companion Seller* Must Do to Be Successful.

1. Use inclusive language
2. Develop a Secrets List
3. Create Backbone spreadsheets
4. Learn to Ballpark on the fly.
5. Treat each project as a series of small conquests.
6. Share information with the client as it arises.
7. Treat the client's money as if it were your own.
8. Develop trust with your client by being trustworthy.

CHAPTER 18

What You Can Expect From Companion Selling.

Companion Selling is a multi-faceted selling and financial strategy developed to change the way your business operates. It requires managerial initiative to succeed, plus a dose of hard work.

Some of the concepts inherent in *Companion Selling* may be foreign to you. Naturally. Our industry operates in a modified retail capacity, and that's what you've been told to emulate.

However, the CI industry – as a whole – is fundamentally flawed. Which explains why so many companies struggle to make payroll and why they don't find consistent growth year-to-year. It also explains why they spawn their own competition from within the ranks.

Clients are not being satisfied. They don't understand how the success of an AV installation is predicated upon the coordination of several disciplines – contracting, electrical, drywalling, painting, cabinetmaking, and so on. These are tradespeople who have not had to coordinate with each other and find no benefit in playing nice. Clients don't understand how aesthetics dictates performance. and how performance dictates price. They don't understand why the integrator has to show up dozens of times over the course of the project, sometimes to check, sometimes to undertake an hour's work. No other subcontractor is subject to these variables – no one except the interior designer.

Which begs the question, "Why is the interior designer successful?" And leads to this answer, "because they employ their own version of *Companion Selling*."

Should you follow the entire *Companion Selling* approach, you will find these results:

A higher close ratio.

Companion Selling techniques allow the prospect to hire you with enthusiasm. So they will. Gone is the antagonistic relationship, replaced with a process of cooperative planning.

A happier staff.

When you commit to mass customization – that is, defining 90% of your systems internally and allowing 10% to be modified by the client – you reduce the variables your staff has to contend with. Your project manager can pop out a Documentation Package quickly. Your programmer will have the bulk of the code sitting on a shelf. Your techs will know the products inside and out. Now your personnel can concentrate on the creative aspects of the business – finding novel ways to integrate audio, video and control into your clients' homes.

Greater profit.

With the modified T&M billing process, you will see significant increases in profit. Labor efficiency (and labor sales) will increase as you capture each labor hour expended. Product sales will rise as you switch from a feature/benefit approach to an aesthetic/functionality approach.

An ability to grow your business.

Many CI companies are stuck at a sales plateau. If you analyze these plateaus, you'll find they're a function of sales hours. One salesperson, using a traditional selling technique, generates about $1m in sales per year, selling 15 discrete projects. That's an average of $67k per project. How can this salesperson double his business? Either by making 30 sales per year – or by selling $130k average projects.

Selling 30 projects per year is just not feasible; there aren't enough hours available in a day. But *Companion Selling* allows you to sell $130k projects with as much effort as a $67k project. *Companion Selling* techniques are transferable. You can grow your business by hiring more salespeople and training them in the techniques. Since the techniques are well defined, you'll find your new salespeople hit the ground running, and are able to generate larger sales quickly.

An exit strategy.

You can't sell your company unless it runs in a predictable fashion. You have to show year-over-year growth in revenue and profits. *Companion Selling* adds predictability and smooths cash flow. That makes investors happy. Because *Companion Selling* is a complete strategy, it can be explained to a potential investor. This adds even more value to your operation.

Financial Benchmark #1: Labor Efficiency.

Here's a piece of AV Solutions' year-end profit and loss statement. There is a revenue line for product sales, and a revenue line for labor sales. (If your P&L statement does not have a labor revenue line, the information is most likely hidden in your accounting software, but aggregated into sales. You can find the actual labor sales figure by digging through your files.)

The labor sales figure represents the amount of labor sold during the year. In this case, $200,000:

	Total
Product Sales	$800,000
Labor Sales	$200,000
Total Sales	$1,000,000

Converting Labor Dollars to Labor Hours.

Let's find out how many labor hours AV Solutions sold in the year. First, we divide labor sales by the average labor selling price. Average labor selling price is your in-house guideline for estimating labor charges.

AV Solutions charges $75/hour for pre-wire and basic install, $100/hour for trim, and $150/hour for programming. Find the weighted average of these selling prices by estimating the number of install hours, trim hours, and programming hours sold during the course of the year.

We'll find AV Solutions' weighted average by looking through its signed proposals for the year. AV Solutions has three technicians each with 2,000 hours available per year. Some programming is done part-time by the project manager. The labor breakdown for the year looks like this:

	Hours Proposed	Selling Rate	Proposed Total	% of Total	Weighted Average
Installation	4,000	$75.00	$300,000	62%	$46.15
Trim	2,000	$100.00	$200,000	31%	$30.77
Programming	500	$150.00	$75,000	8%	$11.54
Total	**6,500**		**$575,000**		**$88.46**

In this example, the average weighted labor selling price per hour is $88.46.

In your case, this might take some work. But it's important to know.

Next, Find the Number of Hours Sold.

When we divide the labor sales by the average weighted labor selling price, we know how many hours of labor AV Solutions sold for the year.

Take the average weighted selling price per hour and divide that into labor sales on the income statement.

Labor Sales: $200,000
Avg. Weighted Selling Price per hour: $88.46

$200,000 / $88.46 = 2,260 Hours Sold

AV Solutions sold 2,260 hours of labor over the course of the year.

Find the Number of Labor Hours Available.

We now know AV Solutions sold and collected money for

2,260 labor hours over the course of the year. But how many labor hours were available? We determine that by going to the payroll records and adding up the number of worker hours logged for the year. According to AV Solutions' payroll records, it paid workers for 6,500 hours.

	Hourly Pay Rate	Annual Hours Paid
Lead Tech	$30	2,000
Junior Tech 1	$20	2,000
Junior Tech 2	$15	2,000
Project Manager	$30	500
Total		**6,500**

Now Find Labor Efficiency.

To find efficiency, divide sold hours by paid hours:

2,260 / 6,500 = 35% efficiency

Yikes. For every hour of labor sold, it took AV Solutions almost three hours to accomplish the work. How do we know this? Because AV Solutions charged its clients for 2,260 hours of labor during the year, but it had 6,500 hours available to sell. That's the number of hours it paid its techs for the year.

Surprised? The average AV company involved in higher-end projects achieves only 30% to 40% labor efficiency. AV Solutions is performing right in the middle of the pack.

Efficiency and Capacity.

AV Solutions sold 2,260 hours. Each technician works 2,000 hours per year. Could one technician have done all the work? Yes, if he worked at 100% efficiency, which is a practical impossibility.

More likely, the guys are busy much of the time, but there are lulls over the course of the year. Maybe no installations take place after holidays. Maybe the workload is slow during certain times of the year and picks up during others. This is the natural cyclicality every service business faces.

Should AV Solutions trim its staff to two techs, they might not be able to accommodate the rush of projects when demand is high. So they've chosen to staff up to meet peak demands – when they need three techs – knowing there'll be cyclical downtime throughout the year.

Staffing up is a capacity issue. AV Solutions made a managerial decision to have the right amount of capacity for peak periods. *Efficiency* is more concerned with a tech's ability to bill for eight hours each day they have eight hours of available time.

Why Efficiency Gains Are "Potential."

Since labor is most likely provided by full-time employees, any small gains in efficiency will likely increase overcapacity and not fall to the bottom line. For AV Solutions, a 1% efficiency gain equates to 3.5 hours per week for the entire install team. What should the company do with these savings? Let the techs go home early?

If AV Solutions is growing, management can use the efficiency gains to grow without having to add another technician so quickly. That 1% efficiency gain increases overcapacity, which permits management to hold off hiring for a longer period of time. AV Solutions saves $5,000 during the "hold-off" period.

If AV Solutions increases efficiencies by 10%, that may be enough to let one of the techs go, while still handling the same amount of business. Those are tangible profits, indeed.

Financial Benchmark #2: Labor Margin.

It's not my intention to delve too deeply into cost based accounting in this book. There are plenty of good reference sources – including your CPA – who can set up your accounting to precisely track product and labor margins. Cost based accounting is a tool financial planners use to get a handle on their costs. Instead of lumping your labor and product expenses together, set up your profit and loss statement to show product revenue and labor revenue as separate accounts. Then, you add all the costs associated with acquiring product and put those into a COGS account called product COGS. You do the same with all your labor costs, including direct labor salaries, benefits, and project related expenses.

Knowing your labor margin requires a bit of manipulation of the expenses on your P&L statement. Typically, most CI companies jumble their labor costs in the expense section of the P&L. We'll need to tease these out of expenses in order to put them in the labor COGS account.

Here's direct labor information derived from AV Solutions:

	Hourly Pay Rate	Annual Hours Paid	Total Direct Labor Salary	Benefits and Taxes	T&E and Automotiv e	Totals
Lead Tech	$30	2,000	$60,000	$6,000	$10,000	$76,000
Junior Tech 1	$20	2,000	$40,000	$4,000	$5,000	$49,000
Junior Tech 2	$15	2,000	$30,000	$3,000	$5,000	$38,000
Project Mgr	$30	500	$15,000	$1,500	$2,000	$18,500
Total		**6,500**	**$145,000**	**$14,500**	**$22,000**	**$181,500**

Let's take all costs associated with direct labor, including salaries, benefits, taxes, vehicle costs, project related travel, and entertainment. Add these up. These are the direct costs of having techs in the field.

Ideally, the top portion of your P&L – including revenue and COGS, will be formatted like AV Solutions'.

We see AV Solutions earns 40 points on product, but only 9 points on labor. Combined, the company earns 34 points of gross margin. These are typical margins for most CI companies.

AV Solutions P&L assuming three technicians.

	Total	
Revenue		
Product Sales	$800,000	
Labor Sales	$200,000	
Total Sales	$1,000,000	
COGS Product		
Total COGS Product	$477,920	40 points
COGS Labor		
Total COGS Labor	$181,500	9 points
Total COGS	$659,420	
Gross Margin	$340,580	34%
Expenses		
Total Expenses	$325,000	
EBITDA	$15,580	2%

Labor Margin is Low. What to Do?

As described previously, labor margins suffer if efficiency is low and overcapacity is high. The easiest way for AV Solutions to decrease labor overcapacity is to eliminate a full-time worker and replace him with a part-time worker or subcontractor as needed.

The following P&L statement restates AV Solutions' year, assuming two technicians handled all the work. Note the changes to labor margin and EBITDA. Note how the labor margin jumps from 9 points to 28 points, and how EBITDA increases 350%.

Is it reasonable to fire a technician and hope the work gets accomplished by the remaining two? Based on labor sales of $200,000 per year, AV Solutions is facing one of two problems. Either it doesn't have enough work for three techs, or it isn't charging for enough labor hours.

AV Solutions P&L assuming two technicians.

	Total	
Revenue		
Product Sales	$800,000	
Labor Sales	$200,000	
Total Sales	$1,000,000	
COGS Product		
Total COGS Product	$477,920	40 points
COGS Labor		
Total COGS Labor	$143,500	28 points
Total COGS	$621,420	
Gross Margin	$378,580	38%
Expenses		
Total Expenses	$325,000	
EBITDA	$53,580	5%

Labor Sales Increase with T&M Billing.

Now we'll begin to see the cash benefits of Companion Selling in action. The following P&L statement shows the affect of T&M billing. In this example, AV Solutions keeps all three techs, yet the modified T&M billing process captures 20 minutes of additional billable labor per day, per technician. Why 20 minutes? Because T&M billing captures all those little charges previously unbilled, as outlined in ***Chapter 13: Milestoning***.

When AV Solutions increases the number of billing hours, it does not increase labor costs. It doesn't increase overhead or expenses. So the income from those extra billable hours flows straight to the bottom line. Earnings have risen to $35,580 which is a 228% increase. That's an enormous gain in earnings,

AV Solutions P&L assuming three technicians and T&M Billing.

	Total	
Revenue		
Product Sales	$800,000	
Labor Sales	$220,000	
Total Sales	$1,020,000	
COGS Product		
Total COGS Product	$477,920	40 points
COGS Labor		
Total COGS Labor	$181,500	28 points
Total COGS	$659,420	
Gross Margin	$360,580	18%
Expenses		
Total Expenses	$325,000	
EBITDA	$35,580	3%

simply by capturing one additional billable hour per day from the install team.

A 10% Increase in Product Revenue.

Now let's focus on product revenue. *Companion Selling*, with its emphasis on relationship building, allows you to sell a better quality solution. And since you're emphasizing aesthetic solutions, you'll find yourself turning to products with better fit and finish. When AV Solutions switches to the *Companion Selling* strategy, its product revenue jumps up 10%. Look how this affects EBITDA.

Product sales are up 10% because AV Solutions is able to present, and get buy-in on higher-quality products. The *Companion Selling* presentation focuses on aesthetics, fit-and-finish, and better performance. These products cost more. For our analysis, product margins remain the same, although in practice they tend to increase.

Labor sales are up because the bi-weekly billing process captures more billable labor. Yet labor costs remain the same.

Expenses don't change with *Companion Selling*, but EBITDA certainly goes up. AV Solutions would earn $67,788 this year following the *Companion Selling* strategy. That's a gain of 435%.

AV Solutions P&L, with *Companion Selling* implemented.

	Total	
Revenue		
Product Sales	$880,000	
Labor Sales	$220,000	
Total Sales	$1,100,000	
COGS Product		
Total COGS Product	$525,712	40 points
COGS Labor		
Total COGS Labor	$181,500	28 points
Total COGS	$707,212	
Gross Margin	$392,788	18%
Expenses		
Total Expenses	$325,000	
EBITDA	$67,788	6%

Using Subcontractors.

Since the CI business has natural variability – meaning some months are strong, others are not – the best financial strategy finds a way to reduce fixed costs and keep expenses variable.

What are fixed costs? These are expenses that pop up every month in a predictable manner. Payroll. Insurance. Vehicle loans. Utilities. These are the checks you write every month regardless of your revenue.

What are your variable costs? To a large degree, the owner's compensation is the only significant variable. Good month = bonus. Bad month = reduced take-home pay.

Now wouldn't it be nice to shift some of those fixed payroll expenses to variable expenses instead? When that tough month comes along, it's most likely a month in which labor sales and income are low. And yet, your fixed labor payroll is constant.

Subcontractors allow you to smooth out your cash flow to match your revenue stream. I'm not limiting my discussion to subcontracted technicians. Think bigger. Think about all of the payroll costs you have, and see if a subcontractor couldn't do the job without adversely affecting your company's performance.

When to Use a Subcontractor.

Value chain analysis – a spiffy marketing theory – holds that you should do the things you do best and subcontract out other aspects of your business. The things you do best should be those elements of your business that define your performance.

I would argue you should keep sales in-house. You should do your own engineering and programming. You should handle your final trim. These elements define your quality level.

Subcontracting the Back Office.

Bookkeeping? I'm not sure why you'd want to do that in-house. It certainly is not an element of your business that defines you in the market. You don't say, "We have the best bookkeeping department in the industry. Our general ledger is pristine."

But most small CI companies handle bookkeeping in-house because it's just another thing the office manager should do. Which begs the question, "What's an office manager for?"

AV Solutions, like most $1m CI businesses, has an office manager. This person has so much excess capacity, he's not quite sure what to do with himself. Here's a time study of AV Solutions' office manager.

AV Solutions Office Manager monthly labor usage.

Task	Hours Required	Instances Per Month	Total Time Required
Invoicing	1	10	10
Collections	1	4	4
Purchase Orders	0.5	12	6
Receipt of PO	0.5	12	6
Entering Bills	0.1	50	5
Payroll	2	4	8
Paying Bills	2	4	8
Answering Phone	2	20	40
Sales Support	1	10	10
Total Hours Needed			97
Hours Available			160
Efficiency			**61%**

The office manager is 61% efficient. He's busy 24 hours each week doing legitimate office work, with the bulk of his time spent answering the phones. The office manager earns $15/hour. Add benefits and taxes, and the cost to maintain this person is $16.50/hour.

It costs AV Solutions $27.22 for every productive hour the office manager works.

Can this person be subcontracted? Absolutely. Because the project manager can pick up the invoicing and purchase order functions – especially if AV Solutions adopts the *Companion Selling* method. A freelance bookkeeper associated with a CPA firm earns about $30/hour, and can handle the bookkeeping functions working about three hours a week. That's $90 each week.

And who will answer the phones? Well, if you do a study of phone calls coming in to a typical CI company, very few of the calls are from clients or prospects. Most are from vendors (calling for payment), techs in the field, and reps. Why would AV

Reallocation of an Office Manager's functions.

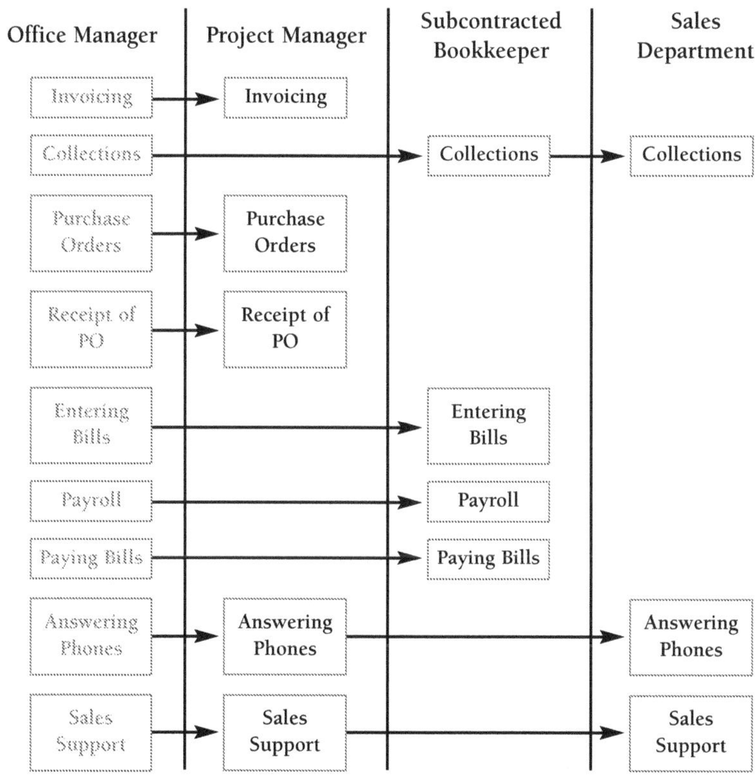

Solutions want to pay the office manager $1,088 each month to field these calls?

Prospects and clients should be given the salesperson's direct phone number. Vendors and techs should be given the project manager's direct line. General calls can be sent to voicemail and be retrieved by management twice a day. That's how you eliminate the role of a receptionist.

Subcontracting Programming.

This is a dicey issue. I'm not a huge fan of subcontracting an aspect of your company as important as programming. And yet, a CI business with sales below $1m cannot justify the expense of a dedicated programmer.

Here's an example where size matters, and subcontractors help keep costs low because utilization is low. The compromise is to subcontract out primary system architecture and programming to a certified programmer, but offer in-house field support for programming as needs arise. That means your project manager or lead tech should have rudimentary programming skills to allow quick fixes, modifications, and service.

A good full-time programmer will cost $115,000 in salary and benefits. So your labor cost per hour for this programmer is $57.50. Let's assume you sell programming at $150/hour. This will require you to sell about 767 hours of programming per year to break even on the programmer's salary. To make 30 points of margin on this programmer, you'll need to sell $164,000 worth of programming per year. On average, programming represents about 10% of the project's total value, so you'd need $1.6m worth of annual revenue in order to have enough programming to keep this person busy 40% of the time.

AV Solutions subcontracts out programming, and relies on its Project Manager to do field work, modifications, simple projects, and service. This is a good use of subcontractors.

Subcontracting Install Labor.

Let's break installation into three categories: prewire, construction-related trim, and final trim. Were you to rate the intrinsic value of each stage, you'd find the closer the project comes to completion, the more important the work. Final trim is critical and prewire is less critical.

Should you choose to subcontract tech labor, you would start with prewire. Perhaps this is subcontracted to the electrician,

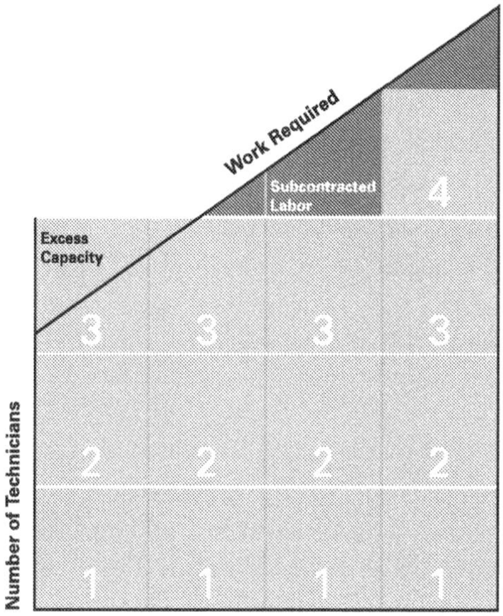

perhaps to a company specializing in structured wiring. In any event, you can increase your techs' efficiencies by removing prewire from their day without reducing the client's confidence in your work.

Reducing or eliminating prewire is a first step in taking advantage of efficiency gains driven by *Companion Selling*. Obviously, as the organization grows, adds employees, and finds changing levels of capacity, the company may choose to handle prewire projects. Prewires are thus treated as a pressure valve for capacity and growth. As the company adds techs, it does the prewires in-house. As the growth consumes the excess capacity, the company subcontracts out prewires until another tech is added, and the cycle continues.

In the above example, AV Solutions has three technicians, but its total available labor exceeds the amount of labor required. The shaded area above the "Work Required" line is excess capacity. As the company grows – and the workload increases, the excess capacity is consumed until the company needs extra help. Instead of hiring another full-time technician and creating excess capacity, AV Solutions chooses to hire subcontractors for prewire projects. The company continues to use subcontractors until it can justify hiring another full-time technician, at which point AV Solutions returns to handling prewires.

Using subcontracted labor in this way keeps in-house labor working to 100% capacity.

Using Permanent Subcontractors.

AV Solutions may choose to keep in-house capacity permanently below the Work Required line, and utilize subcontractors as a permanent part of its business plan. For example, AV Solutions may choose to subcontract all prewires, all the time. In cases where workflow is unseasonably low, the company can handle prewires in-house. When the workload picks up again, it calls on its subcontractors. Using subcontractors in this semi-permanent fashion keeps in-house labor working at full capacity – dramatically increasing labor margin.

About Bay Audio.

Founded in 1999 to provide best-in-class speaker systems to high-end integration companies, Bay Audio has become the industry's leader in strategic partnerships. Bay Audio has extended the integrator's ability to provide exciting, problem-solving speaker systems for home theaters and distributed audio.

Bay Audio products adhere to a strict design goal: We build "apparently better" products that speak of their own value. Elegant, aesthetically mindful of home decor, and acoustically superior, our speaker systems are easy to demonstrate and sell. They allow Bay Audio dealers to achieve the 10% increase in product sales, as outlined in this book.

Learn More About *Companion Selling.*

Bay Audio teaches *Companion Selling* to its dealers through conferences and on-site training. Many companies have found extraordinary profit growth, happier clients, and a more efficient staff by following these concepts. To find out more, call us at 858-486-5040.

www.ingramcontent.com/pod-product-compliance
Ingram Content Group UK Ltd.
Pitfield, Milton Keynes, MK11 3LW, UK
UKHW041851031026
15587UKWH00005B/1147/J

9 780615 158990